HASHTAGPUBLICS

Steve Jones
General Editor

Vol. 103

The Digital Formations series is part of the Peter Lang Media and Communication list.
Every volume is peer reviewed and meets
the highest quality standards for content and production.

PETER LANG
New York • Bern • Frankfurt • Berlin
Brussels • Vienna • Oxford • Warsaw

HASHTAGPUBLICS

The Power and Politics
of Discursive Networks

Edited by Nathan Rambukkana

PETER LANG
New York • Bern • Frankfurt • Berlin
Brussels • Vienna • Oxford • Warsaw

Library of Congress Cataloging-in-Publication Data

Hashtag publics: the power and politics of discursive networks /
edited by Nathan Rambukkana.
pages cm. — (Digital formations; vol. 103)
Includes bibliographical references and index.
1. Online social networks—Political aspects. 2. Information technology—Political aspects.
3. Political participation—Technological innovations. I. Rambukkana, Nathan.
HM742.H3834   302.30285—dc23   2015021776
ISBN 978-1-4331-2899-8 (hardcover)
ISBN 978-1-4331-2898-1 (paperback)
ISBN 978-1-4539-1672-8 (e-book)
ISSN 1526-3169

Bibliographic information published by **Die Deutsche Nationalbibliothek**.
Die Deutsche Nationalbibliothek lists this publication in the "Deutsche
Nationalbibliografie"; detailed bibliographic data are available
on the Internet at http://dnb.d-nb.de/.

Front cover photo: Alina Murad
Back cover: Lois Warwick, 42 Hashtags quilt, as seen on The Quilt Works, Inc.,
based on a pattern from *Fons & Porter's Scrap Quilts* magazine, spring 2014.

The paper in this book meets the guidelines for permanence and durability
of the Committee on Production Guidelines for Book Longevity
of the Council of Library Resources.

Table of Contents

Theorizing Hashtag Publics

Hashtags in Communities, Polities, and Politics

#Acknowledgments

This collection came together in a whirlwind of enthusiasm from the contributors to write about the diverse political uses of the hashtag both inside and outside of Twitter as a platform. First and foremost I want to thank all of the authors, whose insights and research are what make this collection. It was my sincere pleasure and privilege to compile their work and to get to pore over all of their amazing contributions before anyone else. It is a fine honour to watch a field coalesce on your computer screen, and I was provoked and enriched by their insights on the political moments and affordances of hashtags—thanks, all of you, for your incredible and engaging work!

I would also like to thank my colleagues at Wilfrid Laurier. Their professional generosity, camaraderie, and scholarly energy both inspired me to pull this project together and gave me the time and space to pursue it. In particular, my department chairs Andrew Herman and Jonathan Finn helped create and sustain a space that nourished scholarly productivity (sadly, something all too rare in these neoliberal times), and as mentor, Jeremy Hunsinger helped spread the word about this collection and helped us find a publisher.

As a collection of authors, we are very lucky in our choice of publishers and series. Peter Lang has been generous in their support of this project, and we are honoured to be a part of the field-defining Digital Formations series. In particular we would like to thank series editor Steve Jones for all of his insightful comments and guidance, and Mary Savigar and Sophie Appel for keeping us all on track.

Finally, I would like to thank my friends and family, without whose patience and support I could never get anything remotely accomplished, and Zahra, without whom nothing would be worth doing.

#Introduction: Hashtags as Technosocial Events

NATHAN RAMBUKKANA

In "On Actor-Network Theory" (1996), Bruno Latour theorizes around the fundamental ontological nature of things. From his science and technology studies perspective, he explodes the modern understanding of the social sciences as describing some form of pre-existing substance ("the social") and drills down into and unearths their other major aspect, that which looks at the process and becomings of social forms (1996, p. 2). To do this, he draws on chaos theory and event theory, viewing order as emergent, contingent, slices in time of networks of influence in which actants—both human and nonhuman—come together in assemblages that create the *things* we think of as ordered: social structures, technologies, discourses, relationships, movements, concepts, personalities, behaviours, histories, even *matter* (1996, p. 3).

This is a big theory, a top-level theory of everything in which there is "[l]iterally nothing but networks, [and] nothing in between them" (1996, p. 4). This anentropic theory—in which orders are understood as contingent, local and emergent states of a productive and underlying chaos, a "careful plaiting of weak ties" (1996, p. 3) into stronger threads—meshes with understandings as differently situated as M-theory, a development of string theory in which even physical constants such as the speed of light and force due to gravity are understood as contingent and local expressions that might be differently articulated in other universes; and a rhizomatic understanding of social structures, in which social forms are messy and multi-filamented entities that evolve over time and situation, and that, as such, always escape or exceed top-down representations or abstractions (1996, p. 3)—territories that exceed their maps.

In this scheme of understanding, both social networks (in this sense, coming from the 1980s, referring to things such as gangs, unions, subcultures, ideological communities and discourse-cultures) and technical networks (things such as the electrical system, the telephone system) were both overdetermined: "possible *final* and *stabilized* state[s]" of underlying Actor-Networks that might have been ordered otherwise (1996, p. 2; emphasis in original).

This Introduction thinks through the role of hashtags as technosocial events, to use the above understanding of the nature of *things*, networks, and organization to think through how we can understand hashtag-mediated discursive assemblages. What are these networks—ones that we can think of as Actor-Networks comprising actants or actors that are variously individual (people crafting, deploying, amplifying, or utilizing hashtags), technological (the hardware and software that encode specific affordances around these tags through Twitter and other technologies), collective (groups or other collectivities of people), or corporate (institutions, corporations, states)? Are they communities, publics, discourses, discursive formations, *dispositifs*, something else? This collection argues and demonstrates that the media ecology subtended by hashtags can be any of these, as they can be drawn in to articulate with all of these organizational possibilities in different circumstances and configurations.

Eschewing a simple notion of technological determination, this collection is premised on the argument that the form and matter of these assemblages can take on different emergent qualities based on the particular actants that are at work in shaping each tag as its own unique and individual event. The performativity of such utterances makes hashtags discourse that recognizes itself as such (sometimes); an affective amplifier (sometimes); useful in linking or constituting particular publics (sometimes); and even able to subtend communities (sometimes)—depending on how it is deployed. This complexity of possible forms and dynamics taps its insight from Simondon's critique of hylomorphism (or the conceptual division and opposition of form and matter).

In "Forme et Matière" (1995), Gilbert Simondon discusses how *things*—all *things*—take form. It is neither the essential qualities of the underlying matter, nor the shaping qualities of an applied form that dictate the final shape and nature of any material thing. He uses the metaphor of the formation of a brick to illustrate this profound but simple point. Neither the mix of components alone nor the brick mould alone is sufficient to produce a brick. A mould applied to the wrong kind of clay, or with too much liquid, or not enough, or with stones, will not produce the brick as a finished technical form (p. 38). Likewise, a mould with different qualities (size, shape, materials, porousness, rigidity, flexibility) acting on the same clay would not produce the same brick either. It is the combination of these exact conditions and in the right configuration—with the correct amount of time, the proper preparation, the right amount and duration of pressure, the appropriate humidity, the same procedure for filling and emptying the mould—that produces the finished thing: a brick, a piece of worked-on matter than can then be taken up as material to build further forms and structures (such as a building, a wall, a path) (p. 55). In other words, *la prise de forme* of a brick requires both of these and moreover a specific process—an event—to work these elements together into their ultimate shape.

According to Brian Massumi (2004), Simondon's writing on form and matter can be usefully mobilized to think through how discourse forms and circulates. Rather than a simplistic reading that would see discourse as a mimetic reflection of human culture or a deterministic one that would see it as a top-down shaper of culture, Massumi's mobilization posits discourse as technosocial event, shaping and shaped, *forme et matière*. It is the complex singularity that gains substance through its ongoing becoming; it is both medium and message.

Hashtags, as discursive assemblages, could be rendered similarly. Both text and metatext, tag and subject matter, pragmatic and metapragmatic speech act (Benovitz, 2010), hashtag-mediated discursive assemblages are neither simply the reflection of pre-existing discourse formations nor do they create them out of digital aether. Rather, they are nodes in the becoming of distributed discussions in which their very materiality as performative utterances (Sauter & Bruns, Chapter 3, this volume) is deeply implicated.

For example, there has been much discussion about the role of hashtags, as well as Twitter, Facebook, and social media broadly, in the Arab Uprisings (e.g., Papacharissi & Oliveira, 2012; Douai, 2013). While some in the public sphere go so far as to call such techno-implicated events or movements, especially with respect to Egypt, "social media revolutions," or "the Facebook/Twitter revolution," others are more muted in their attribution of causal force to these technologies.

In Aziz Douai's article "'Seeds of Change' in Tahrir Square and Beyond" (2013), he picks apart and reframes this notion. Looking at the larger technological and human contexts of movements in and transformations of Arab societies, he shows that the technological intervention of social media activism was only the most recent iteration of concatenated changes that have been at work on the shapes of connections, knowledge production, and information circulation in these societies since satellite television started to open up these public spheres in the 1960s. Digital and other technologies, he argues, have long articulated with Arabs generally and Egyptians specifically forging a vibrant public sphere. Using any and all of videotapes and DVDs, CD-ROMs, computer media, digital still and video cameras, video games, the Internet, and the rich convergence of all media the Internet now provides (e-mail, messaging, blogging, archiving and distributions of texts, image and video sharing, video chat, and SNSs such as Facebook,

Instagram, and Twitter), Arabs have spoken back to both state and world power, worked to expose abuses, and commented on authoritarian regimes—their own, those of neighbouring countries, and those of places such as the U.S., Europe, and Canada. He argues that, given this, not just social networking technologies but *all* such technologies can be argued to be potential technological seeds of change (Douai, 2013, p. 25).

At the same time, however, he argues that they would not have been put to these uses (or—to start to steer this in the direction we are interested in theoretically—they would not have been propelled to *act* in such ways) without a vibrant counterpublic, robust social movements, and dedicated individuals (Douai, 2013, p. 30) that comprise a parallel influence, the "social seeds of change." These two sets of gametes cross-pollinate and influence each other—new technologies create new affordances that inspire new politics, and political uses of technologies emerge from the influence of the political imagination on a technical capability: for example, we can see both at work simultaneously in how citizen-captured video of police brutality gets spread far and wide through the technoscape (shared via YouTube, Twitter, Facebook, blogs, etc.). In this example, we can return to Latour and Simondon both. The questions of "Which moves what?" becomes moot in an Actor-Network framework which would see both the technologies and the individuals as actors working to influence each other and articulate together, and Simondon would see the matter (protest of state) and form (social media) as implicated in the same event, conditioning each other in the singular encounter.

Returning to hashtags, and what they can specifically engender as actants and actors in the networks they prehend to, we can see them as pathways to an open and non-predefined set of communicative encounters and architectures, a crossroads between form and matter, medium and message entangled (see also Rambukkana, Chapter 2, this volume).

On a continuum with a derided notion of "hashtag activism" at one pole and an inflated notion of the power of "hashtag uprisings" at the other are all political hashtags and all the publics they subtend. And it truly is a continuum, of which this collection offers only a modest sampling—a snapshot, a section. Ranging from hashtags such as **#isthenipplepolitical** to **#tahrir**, from **#winning** to **#BlackLivesMatter**, from **#FirstWorldProblems** to **#rapedneverreported**, the hashtag-mediated public sphere is not one thing and resists any singular characterization. To inscribe some hashtags into the world, to instance them in ink or through digital data, could get you killed; others could get you ostracized from social spaces you frequent, or get you fired; others might do nothing whatsoever. They are a technic (which is to say, both a technique and a technology) of the social, and in their performativity are events that map together and encompass not just the tag itself but the network of human and nonhuman actors that come together in such configurations: tags, technologies, taggers, conversations, press coverage in other media.

As Axel Bruns and Jean Burgess argue (2011, p. 7, and Chapter 1, this volume), "To include a hashtag in one's tweet is a performative statement: it brings the hashtag into being at the very moment that it is first articulated, and—as the tweet is instantly disseminated to all the sender's followers—announces its existence." Which is to say that it is a saying which is also a doing, to paraphrase Austin (1975, p. 5), an utterance that is at the same time an action. To use a hashtag is to imbue that actant with the properties of material action, to move it from the virtual to the actual, to make it an *actor* in its own right. To accelerate it into motion in this way, to imbue it with affect, allows it to affect in turn; it weaves it into Actor-Networks of potential new configuration. That said, Bruns and Burgess note that once this performative act happens, "the extent to which the community around the hashtag becomes more than an issue public of one depends on its subsequent use by other participants" (2011, p. 7, and Chapter 1, this volume). I agree, with two provisos. The first is that while it may be a community or a public that forms around a given tag, these are only two possibilities among many. Other configurations to which hashtags are articulated are possible—to pick a few at random: advertising campaigns, political platforms, social movements, smear campaigns, activist protests, harassment crusades,[1] consumer products, and revolutions (see Figure 1.1). Not all hashtags have politics, create publics, or maintain communities, in other words. But some can, and some do. The obverse is also true: just because one hashtag does not articulate to a greater movement, community, or politics, it does not mean another might not. A hollow hashtag public around a tag such as **#yolo** does not negate the force around **#Ferguson** or **#outcry** (see Rambukkana, Chapter 2, and Antinakis-Nashif, Chapter 7, this volume).

As with Simondon's brick that inherits its singular nature neither from the clay poured nor from the mould applied but from both—from the *event* of their combination, the process of their application to each other, the sequence and specific method of their admixture—hashtag-mediated assemblages inherit their character neither solely from the social material poured into them nor from their specific nature as technology or code alone but from the singular composition of each tag in their "continuous locality" (Latour, 1996, p. 6): in how they are crafted; for what purpose; among what other actants and actors, individual, technical, communal or corporate; as well as in what spaces, through what technologies, with what coverage; and finally articulated with which effects, with what temporality, and through which history. Hashtags, as a form of digital intimacy, are a way that things in the world touch other things in the world and form networks with them; they are multiple, open-ended, and contingent phenomena.

Figure 1.1. Hashtags on consumer products (image credit: Alina Murad, 2015).

BREAKDOWN OF SECTIONS

The first section of the collection, "Theorizing Hashtag Publics," takes a panoramic view of some of the theoretical issues that attend the intersection of hashtags and the political. Covering governmental politics, industrial politics, the politics of representation, and the temporality of discourse, these papers run the gamut of levels of hashtag engagement and explore the deep theory and history of the hashtag.

The section begins with an updated version of Axel Bruns and Jean Burgess's field-defining study on hashtags as "ad hoc publics" that goes on to consider the increasingly calculated nature of the publics that hashtags are conceived to collect. With public reflections on Australian leadership spills as an object, this chapter—cited by the vast majority of the other chapters in the collection—defined the possible contours of hashtag community and public sphere engagement and, in this new version, complicates the picture with a read on where algorithmic logic and curated and commodified hashtags bring us. Nathan Rambukkana's paper further theorizes the political potential of hashtag publics, focusing on loud and angry protest publics such as **#RaceFail** and **#Ferguson**. He looks at how the energy of race-activist hashtags can spill over into other mediated spheres and other publics, affecting popular culture representations, the publishing and broadcasting industries, fan cultures, and discourse about race and representation broadly. Theresa Sauter and Axel Bruns continue the top-down investigation of the politics of hashtags through exploring **#auspol**, the major meta-hashtag appended to discussions of Australian politics. Investigating who uses this tag and how the discussions draw users into a particular kind of public sphere engagement—one that is even observed by some as an aggregate *vox Twitteratorum*—is informative for looking at hashtags as a political technology broadly. From top-down Australian politics, we move to bottom-up Australian politics, with Jean Burgess, Anne Galloway, and Theresa Sauter's exploration of the uses of **#agchatoz**, a tag used by Australian farmers and interested parties discussing agricultural issues. Tags such as this and similar ones for other regions create the possibility of broad discursive channels for discussing the grounded and embodied politics of particular professions both within and across geographical regions. In the last paper in this section, Daniel Faltesek does a deep reading of the temporality of the hashtag, leaving us with a series of important questions: How do hashtags create a sense of sequence, of event, of history? How can we distinguish true hashtag publics from the "false publics" of neoliberalism and political rhetoric? How might thinking of hashtagged discourse as unmediated by technology and covert intentionalities be dangerous?

The second section, "Hashtags and Activist Publics," explores how hashtag use is mobilized for activist causes. What affordances do hashtags lend to activism—on the Internet, in the media, and even on the street? But also, how might skirmishes for the meaning of hashtags, or between competing hashtags, create interference or even pitched battles for the attention of others on particular issues?

Aaron S. Veenstra, Narayanan Iyer, Wenjing Xie, Benjamin A. Lyons, Chang Sup Park, and Yang Feng use quantitative analysis to investigate retweeting behaviour with respect to the 2011 Wisconsin labour protests. While tag channels were useful for spreading information and mobilizing for local events, they were less useful at building solidarity across other locations, showing how social media organizing cannot be a comprehensive replacement for traditional social movements and networks. In the European context, Anna Antonakis-Nashif explores how a feminist hashtag, **#aufschrei** (outcry), acted as a catalyst for a sharpened societal debate of sexism in Germany, attracting a powerful but ambivalent publicity that at once raised this issue to the level of mass-societal discussion but in doing so risked reducing it to a singular prominent incident and invited an antifeminist backlash. In South America, Carlos d'Andréa, Geane Alzamora, and Joana Ziller consider how hashtags have an "intermedia agency" that spills from the Internet to the street and back again in protests about the FIFA World Cup 2014 in Brazil.

They mobilize the concept of a "hashtag war" to think through how similar, opposing tags and slogans were used by protestors, the government, and sponsors to wrestle for control of the public sphere understanding of this mega sporting event. Back in the U.S. context, Jenny Ungbha Korn looks at temporary activist publics around Proposition 8, and how they move from queer anger to queer empowerment. Thinking beyond a limited notion of a virtual community that tapers off as a "failed" one, she looks at how a victorious hashtag protest public, such as that around **#FuckProp8**, might speak to the natural life cycle of tags, making us rethink how we understand online communities that are "in stasis." Finally, Stacy Blasiola, Yoonmo Sang, and Weiai Wayne Xu investigate the technics of hashtag publics and how activists use the technical affordances of the medium to further their causes. In their analysis of how opponents of the Cyber Intelligence Sharing and Protection Act (CISPA) use **#CISPA**, they pick apart how this channel is variously used for communicative and technical actions.

The third section, "Art, Craft, and Pop Culture Hashtag Publics," looks at how cultures of fandom and of creative practice use hashtags to further their intensive pursuits. In all of these chapters, the deep intermedia nature and interpenetration of fandom, art, and digital culture is apparent.

Commenting on the space between "realness" and "**#realness**" in the layered texts of *Paris Is Burning*, *RuPaul's Drag Race*, and the art of Wu Tsang, Andy Campbell gives a richly textured account of how the hashtagification of countercultural lingo can sometimes lead to it losing its critical edge and specificity through mass social media uptake. Failed hashtags, appropriated hashtags, and the tensions between hashtags and countercultural publics are all discussed in this piece. Moving from pop cultures to craft cultures, Amanda Grace Sikarskie looks at quilting communities on Instagram and what it means to live the **#Quilt Life**. How do older craft cultures use new media to connect and communicate, and how does that put them into (sometimes surprising) dialogue with other communities, such as fan cultures that might share a tag at random or have fannish interests that are expressed through craft? Focusing on a different kind of popular hashtag uptake (this one more problematic), Andrew Peck's investigation of the disparate politics and tensions among those posting with the tag **#firstworldproblems** shows how a parody of Twitter and its privileged users, forged outside of the medium itself, can be recirculated within it to diminishing returns on its politics. In this case, the sharpest political publics of this hashtag seem to be, ironically, the ones formed outside of its technical tagging function, on sites such as SomethingAwful.com and YouTube. In the last paper of this section, Anthony Santoro explores the superdiversity of sports fan communities in relation to **#RaiderNation** and how this online and offline community both contests and is complicit with neoliberalism in its intensive and devoted fannish pursuits.

Finally, the fourth section, "Hashtags in Communities, Polities, and Politics," looks at how more traditional polities such as racial groups, language communities, ideological collectivities, and even academics use hashtags in ways that variously express, cater to, and facilitate their collective identities.

Meredith Clark's and Nia Cantey and Cara Robinson's two chapters both deal with the phenomenon of **#BlackTwitter**. Through interviews and analyses of tweets, Clark investigates multiple scandals that "broke" on Black Twitter and how they played out through the use of various hashtags, as well as how Black Twitter works as a meta-network that serves multiple communicative goals for both its users and mainstream sources "listening in" on these important cultural conversations. Cantey and Robinson take a different approach, meticulously picking apart what Black Twitter is and is not, investigating not only the counterpublic power of this hashtag public but also whose voices it might privilege and whose it might exclude. Next, Magdalena Olszanowski strikes out beyond Twitter to investigate how hashtag use on Instagram has organized an intimate public of women who engage in feminist exchange through image sharing. Their communities are vital but fraught, important but at risk due to censorship and the threat of account deletion, where hashtags can be double-edged swords that are useful for finding "like-minded people" but also mark their artistic creations as more visible to those who might flag them as "problematic" or have the power to erase them. Sylvain Rocheleau and Mélanie Millette return to Twitter to investigate how appending meta-hashtags and using hashtag co-occurrence in posts can tie together linguistic and regional linguistic communities, with French Canadian minority language communities as an example. Brett Bergie and Jaigris Hodson's paper is a crucial counterpoint to many of the other papers in the collection. They argue in their analysis of Alberta, Canada budget hashtag use that there was little actual discussion over the tag channels, as they were dominated by tweets from traditional media sources that not only flooded the channel with mass media content but also fragmented it, pulling attention back to their own walled, ad-driven discussion forums and media spaces. Finally, sava saheli singh turns a reflexive eye on academics' use of hashtags to help organize and share the business of academe, including their performative uses for showing membership in academic subcommunities such as associations, or showing scholarly interest in specific topics, as well as how academics mine hashtags for data, use them in classrooms, and deploy them to create lively backchannels or even "parallel signaling streams" at conferences.

Together, this collection considers hashtag use across multiple continents and media—both digital and nondigital. While it is limited by mostly Western sources, and as such can be seen only as an entry point for a broader discussion of hashtag publics, the diversity of its subject matter speaks to the networks teased together by these technical actors and those that accelerate them into action. Cutting across sections, the chapters collected here discuss issues surrounding gender, race, class, and sexuality; surveillance, neoliberalism, industry, privilege, and language rights; fandom, academia, art, craft, politics, and pop culture. Hashtags can help form communities, collect publics, incite protest, inform polities, and even hawk products. They are equally available to progressives fighting power and to the powerful themselves. In their current popularity, they manifest across media and are mobilized for both the political and the trivial. Rough, emergent, hard to tame or pin down, it remains to be seen if hashtag use will fizzle out with the introduction of new "it" technologies or transform anew to trend into the new spaces of the communication landscape.

NOTE

1. Harassment crusades are larger, more organized attempts to attack people or groups over the Internet, such as **#GamerGate** (see also Rambukkana, Chapter 2, this volume).

REFERENCES

Austin, J. L. (1975). *How to do things with words*. Cambridge, MA: Harvard University Press.

Benovitz, M. G. (2010). "Because there aren't enough spoons": Creating contextually-organized argument through reconstruction. In *National Communication Association Annual Conference, Conference Proceedings* (pp. 124–130). Washington, DC: National Communication Association.

Bruns, A., & Burgess, J. (2011). *The use of Twitter hashtags in the formation of ad hoc publics*. Paper presented at 6th European Consortium for Political Research General Conference, August 25–27, University of Iceland, Reykjavik. Retrieved from http://eprints.qut.edu.au/46515/

Douai, A. (2013). "Seeds of change" in Tahrir Square and beyond: People power or technological convergence? *American Communication Journal, 15*(1), 24–33.

Latour, B. (1996). On actor-network theory. A few clarifications plus more than a few complications. *Soziale Welt, 47*, 369–381.

Massumi, B. (2004). *Analyse des discourse et des messages médiatisés*. Ph.D. seminar, Université de Montréal.

Papacharissi, Z., & Oliveira, M. de F. (2012). Affective news and networked publics: The rhythms of news storytelling on #Egypt. *Journal of Communication, 62*, 266–282.

Simondon, G. (1995). Forme et matière. In *L'individu et sa genèse physico-biologique* (pp. 37–64). Grenoble, France: Millon.

Theorizing Hashtag Publics

CHAPTER ONE

Twitter Hashtags from Ad Hoc to Calculated Publics[1]

AXEL BRUNS AND JEAN BURGESS

INTRODUCTION

From its early beginnings as an instant messaging platform for contained social networks, Twitter's userbase and therefore its range of uses increased rapidly. The use of Twitter to coordinate political discussion, or crisis communication especially, has been a key to its legitimisation, or 'debanalisation' (Rogers, 2013), and with the increased legitimacy has come increased journalistic and academic attention—in both cases, it is the hashtag that has been perceived as the 'killer app' for Twitter's role as a platform for the emergence of publics, where publics are understood as being formed, re-formed, and coordinated via dynamic networks of communication and social connectivity organised primarily around issues or events rather than pre-existing social groups (cf. Marres, 2012; Warner, 2005).

The central role of the hashtag in coordinating publics has been evident in contexts ranging from general political discussion through local, state and national elections (such as in the 2010 and 2013 Australian elections) to protests and other activist mobilisations (for example, in the Arab Spring as well as in Occupy and similar movements). Twitter hashtags have also featured significantly in other topical discussions, from audiences following specific live and televised sporting and entertainment events to memes, in-jokes and of course the now banal practice of live-tweeting academic conferences.[2]

Research into the use of Twitter in such contexts has also developed rapidly, aided by substantial advancements in quantitative and qualitative methodologies for capturing, processing, analysing and visualising Twitter updates by large groups of users. Recent work has especially highlighted the role of the Twitter hashtag as a means of coordinating a distributed discussion among large numbers of users, who do not need to be connected through existing follower networks.

Twitter hashtags—such as **#ausvotes** for the 2010 and 2013 Australian elections, **#londonriots** for the coordination of information and political debates around the 2011 unrest in London, or **#wikileaks** for the controversies around Wikileaks—thus aid the formation of ad hoc publics around specific themes and topics. They emerge from within the Twitter community—sometimes as a result of preplanning or quickly reached consensus, sometimes through protracted debate about what the appropriate hashtag for an event or topic should be (which may also lead to the formation of competing publics using different hashtags). But hashtag practices are therefore also far from static and may change over time: the prominent role in organising ad hoc discussion communities which existing studies have ascribed to hashtags may now be changing as users are beginning to construct such communities through different means, and as Twitter as a platform curates and mediates hashtags more actively.

BACKDROP

Australia, 23 June 2010: rumours begin to circulate that parliamentarians in the ruling Australian Labor Party (ALP) are preparing to move against their leader, Prime Minister Kevin Rudd. Rudd was elected in a landslide in November 2007, ending an 11-year reign by the conservative coalition, but his personal approval rates have slumped over the past months, further fuelling his colleagues' misgivings over his aloof, bureaucratic leadership style. In spite of the fact that opinion polls continue to predict a clear victory for the ALP in the upcoming federal elections later that year, that Wednesday evening, Labor members of parliament are considering the unprecedented—the replacement of a first-term prime minister, barely 2½ years after his election.

As rumours of a palace revolution grow, Australia's news media also begin to cover the story—special bulletins and breaking news inserts interrupt regular scheduled programming. Amongst the key spaces for political discussion that evening is Twitter: here, those in the know and those who want to know meet to exchange gossip, commentary, links to news updates and press releases, and photos of the gathering media throng. The growing crowd of Twitter users debating the impending leadership spill includes government and opposition politicians, journalists, celebrities, well-known Twitter micro-celebrities, and regular users; by midnight, some 11,800 Twitter users will have made contributions to the discussion.[3]

Events such as this demonstrate the importance which Twitter now has in covering breaking news and major crises; from the killing of Osama bin Laden through the Sendai earthquake and tsunami to the disappearance of Malaysia Airlines flights MH370 and MH17, Twitter has played a major role in covering and commenting on such events. The most widely recognised mechanism for the coordination of such coverage is the hashtag: a largely user-generated mechanism for tagging and collating those tweets which are related to a specific topic. Senders include hashtags in their messages to mark them as addressing particular themes. For Twitter users, following and posting to a hashtag conversation makes it possible for them to communicate with a community of interest around the hashtag topic without needing to go through the process of establishing a mutual follower/followee relationship with all or any of the other participants; in fact, it is even possible to follow the stream of messages containing a given hashtag without becoming a registered Twitter user, and these days, a curated version of the hashtag stream may even be broadcast alongside television news coverage or displayed on a public screen. In its potentially network-wide reach, hashtagged discussion operates at the macro level of Twitter communication (Bruns & Moe, 2014), compared to the structurally more insular exchanges through the personal publics (Schmidt, 2014) of follower networks at the meso level, and to targeted public @replying between individual users at the micro level.

In the case of the ALP leadership challenge, Twitter users quickly settled on the hashtag **#spill** (Australian political slang for a party room vote on the leadership); during 23 June 2010 alone, the 11,800 participating Twitter users generated over 50,000 tweets containing **#spill**—not particularly large numbers by today's standards, but a dramatic spike at the time. The majority of those tweets are concentrated between 19:00 and midnight, as the rumours were further amplified by mainstream media coverage; between 22:00 and 23:00, **#spill** tweets peaked at more than 4,500 per hour (or 75 per minute), while activity prior to 19:00 barely reaches 10 tweets per hour (Bruns, 2010a, 2010b). This fast ramping up of activity in the evening also demonstrates Twitter's ability to respond rapidly to breaking news—an ability which builds not least on the fact that new hashtags can be created ad hoc, by users themselves, without any need to seek approval from Twitter administrators. As we will argue in this chapter, this enables hashtags to be used for the rapid formation of ad hoc issue publics, gathering to discuss breaking news and other acute events (Burgess, 2010; Burgess & Crawford, 2011).

But not all hashtags are topical, and not all topical hashtags are used to facilitate the gathering of ad hoc publics. As preferences for the use of Twitter as a social medium change over time, and as the hashtag is employed and appropriated for various strategic and tactical means by a range of stakeholders, it is possible that the role of the hashtag in public communication is also changing. This may cause it to lose its utility in coordinating topical public discussion on an ad hoc basis.

A SHORT HISTORY OF THE TWITTER HASHTAG

In the early phases of adoption following its launch in 2006, Twitter had almost none of the extended functionality that it does today. Twitter users were invited to answer the question 'What are you doing?' in 140 characters or less, to follow the accounts of their friends, and little else (see Burgess, 2014). Many of the technical affordances and cultural applications of Twitter that make its role in public communication so significant were originally user-led innovations, only later being integrated into the architecture of the system by Twitter, Inc. Such innovations include the cross-referencing functionality of the @reply format for addressing or mentioning fellow users, the integration of multimedia uploads into Twitter clients and—most significantly for this paper—the idea of the hashtag as a means to coordinate Twitter conversations.

As a concept, the hashtag has its genealogy in both IRC channels and the Web 2.0 phenomenon of user-generated tagging systems, or 'folksonomies', common across various user-created content platforms by 2007, with Flickr and del.icio.us being the most celebrated examples. The use of hashtags in Twitter was originally proposed in mid-2007 by San Francisco–based technologist Chris Messina, both on Twitter itself and in a post on his personal blog, titled 'Groups for Twitter, or a Proposal for Twitter Tag Channels' (Messina, 2007a). Messina called his idea a 'rather messy proposal' for 'improving *contextualization, content filtering* and *exploratory serendipity* within Twitter' by creating a system of 'channel tags' using the pound or hash (#) symbol, allowing people to follow and contribute to conversations on particular topics of interest. The original idea, as the title of Messina's post indicates, was linked to proposals within the Twitter community for the formation of Twitter user groups based on interests or relationships; counter to which Messina argued that he was 'more interested in simply having a better eavesdropping experience on Twitter'. So rather than 'groups', hashtags would create *ad hoc channels* (corresponding to IRC channels) to which groupings of users could pay selective attention. While Messina went on to propose complex layers of user command syntax that could be used to manage and control these 'tag channels' (including subscription, following, muting and blocking options), the basic communicative affordance of the Twitter hashtag as we know it today is captured in his vision for the 'channel tag':

> Every time someone uses a *channel tag* to mark a status, not only do we know something specific about that status, but others can *eavesdrop* on the context of it and then join in the

channel and contribute as well. Rather than trying to ping-pong discussion between one or more individuals with daisy-chained @replies, using a simple #reply means that people not in the @reply queue will be able to follow along, as people do with Flickr or Delicious tags. Furthermore, topics that enter into existing channels will become visible to those who have previously joined in the discussion. (Messina, 2007a)

At first there was little take-up of Messina's idea—until the October 2007 San Diego bushfires demonstrated a clear use-case (and partly as a result of Messina's activism during that event, urging people to use the hashtag to coordinate information—see Messina, 2007b). Over time, the practice became embedded both in the social and communicative habits of the Twitter user community and in the architecture of the system itself, with the internal cross-referencing of hashtags into search results and trending topics. Of course, like most successful innovations, the hashtag's original intended meaning as an 'invention' has long since become subverted and exceeded through popular use; this is largely attributable to its stripped-down simplicity and the absence of top-down usage regulation—there is no limit or classification system for Twitter hashtags, so all a user needs do to create or reference one is to type the pound/hash symbol followed by any string of alphanumeric characters. In the years since 2007, through widespread community use and adaptation, the hashtag has proven itself to be extraordinarily high in its capacity for 'cultural generativity' (Burgess, 2012) and has seen a proliferation of applications and permutations across millions of individual instances—ranging from the coordination of emergency relief (Hughes & Palen, 2009) through the most playful or expressive applications, as in Twitter 'memes' or jokes (Huang, Thornton, & Efthimiadis, 2010), to the co-watching of (and commentary on) popular television programs (Deller, 2011; Highfield, Harrington, & Bruns, 2013) and the coordination of ad hoc issue publics, particularly in relation to formal and informal politics (Small, 2011).

THE USES OF HASHTAGS

While the focus of this chapter is on the use of hashtags to coordinate public discussion and information-sharing on news and political topics, it is useful to outline a brief typology of different hashtag uses.

In the first place, hashtags can be used to mark tweets that are relevant to specific known themes and topics; we have already encountered this in the example of the Australian leadership #spill. Here, a drawback of the ad hoc and uncoordinated emergence of hashtags is that competing hashtags may emerge in different regions of the Twittersphere (for example, #eqnz as well as #nzeq for coverage of the Christchurch, New Zealand earthquakes in 2010 and 2011), or that the same hashtag may be used for vastly different events taking place simultaneously (for instance, **#spill** for the BP oil spill in the Gulf of Mexico during the first half of 2010, as well as for the leadership challenge in the Australian Labor Party in June of the same year).

Twitter users themselves will often work to resolve such conflicts quickly as soon as they have been identified—and such efforts also demonstrate the importance of hashtags as coordinating mechanisms: users will actively work to keep 'their' hashtag free of unwanted or irrelevant distractions, and to maximise the reach of the preferred hashtag to all users. Where—as in the case of **#spill**—both sides have a legitimate claim to using the hashtag, it is often the more populous group which will win out; on 23 June 2010, for example, the political crisis in Australia drew considerably more commenters than the Gulf of Mexico oil spill which had been in the news for several months already, and suggestions to disambiguate the two by marking leadership-related posts with alternative hashtags such as **#laborspill**, **#spill2**, or **#ruddroll** were not widely heeded. Instead, Australian Twitter users occasionally posted messages to explain the takeover of 'their' hashtag to those still following the oil spill—for example:

> those who do not know a **#spill** in leadership terms is basically saying the big job is now vacant no relation to the oil spill of bp style

while those following from outside Australia expressed their confusion at this sudden influx of new messages:

> Ok what's with the **#spill** tag? Has BP dumped more oil?

On the other hand, where—as in the case of **#eqnz** versus **#nzeq**—what should be a unified conversation is splintered across two or more hashtags, participants often try to intervene to guide more users over to what they perceive to be the preferred option. Here, messages from major, authoritative accounts can act as influential role models for 'correct' hashtag use, but users will also encourage those authorities to use hashtags 'properly' if they do not do so initially:

> @NZcivildefence please use **#eqnz** hashtag. Thanks.

At the same time, a splintering of conversations may also be desirable as themes shift or diversify. So, for example, while general discussion of everyday political events in Australia is commonly conducted under the hashtag **#auspol**,[4] separate hashtags are regularly used to track parliamentary debate during Question Time (**#qt**) or to comment on the weekly politics talk show *Q&A* on ABC TV (**#qanda**), as well as for the discussion of specific issues or crises. Where sensible, or where they wish to maximise their message's reach, Twitter users may also use multiple hashtags to address these various, overlapping constituencies.[5]

Such examples, therefore, underline the interpretation of using a thematic hashtag in one's tweet as an explicit attempt to address an imagined community of users who are following and discussing a specific topic—and the network of Twitter users which is formed from this shared communicative practice must be understood as separate from follower/followee networks. At the same time, the two network layers overlap: all public tweets marked with a specific hashtag will be visible *both* to the user's established followers *and* to anyone else following the hashtag conversation. Users from the follower network who respond and themselves include the hashtag in their tweets thereby also become part of the hashtag community, if only temporarily, while responses to or retweets of material from the hashtag conversation are also visible to the follower network. Similarly, some users may retweet topical tweets from their followers while adding a hashtag in the process, thereby making those tweets visible to the hashtag community as well. Each user participating in a hashtag conversation therefore has the potential to act as a bridge between the hashtag community and their own follower network (cf. Bruns & Moe, 2014).

At the same time, not all users posting *to* a hashtag conversation also *follow* that conversation itself: they may include a topical hashtag to make their tweets visible to others following the hashtag, thereby increasing its potential exposure, but may themselves continue to focus only on tweets coming in from their established network of followees (this is especially likely for very high-volume hashtag streams). Conversely, not all relevant conversations following on from hashtagged tweets will themselves carry the hashtag: to hashtag a response to a previous hashtagged tweet, in fact, may be seen as *performing* the conversation in front of a wider audience, by comparison with the more limited visibility which a non-hashtagged response would have.

Beyond thematic, topically focussed uses of hashtags, a number of other practices are also evident. A looser interpretation of hashtagging is present in tweets which simply prepend the hash symbol in front of selected keywords in the tweet:

#japan #tsunami is the real killer. #sendai #earthquake

PGA only 0.82g. 2011 #chch #eqnz 2.2g http://j.mp/ecy39r

Such uses may be a sign that hashtags for breaking events have not yet settled (and that the sender is including multiple potential hashtags in their message in order to ensure that it is visible to the largest possible audience), or that the sender is simply unaware of how to effectively target their message to the appropriate community of followers—additionally, of course, they could also be read as a form of Twitter spam. For the most part, at any rate, it is unlikely that significant, unified communities of interest will exist around generic hashtags such as **#Japan** or **#Australia**, for example: outside of major crises affecting these countries (when we may reasonably expect the vast majority of tweets to refer to current events), tweets carrying such generic hashtags will cover so wide and random a range of topics as to have very little in common with one another.

An alternative explanation for the use of such generic hashtags, then, is as a simple means of emphasis—especially in the absence of other visual means that may be used to embellish tweets (such as bold or italic font styles). A hashtag like **#Australia**, therefore, should usually be seen as equivalent to text decorations such as '_Australia_' or '*Australia*', rather than as a deliberate attempt to address an imagined community of Twitter users following the **#Australia** hashtag conversation, such as it may be.

Such emphatic uses are especially evident in hashtags which (often ironically) express the sender's emotional or other responses—for example, **#sigh**, **#facepalm**, or **#headdesk**. Here, hashtags take on many of the qualities of emoticons like ';-)' or ':-O'—they are used to convey extratextual meaning, in a Twitter-specific style. Additionally, however, some of these hashtags—for example, **#firstworldproblems** or **#fail**[6]—have also morphed into standing Twitter memes, to the point where some users may in fact have started to follow them for the entertainment they provide; here, a community of interest of sorts may once again have formed, then, even if few of the hashtagged messages themselves are intentionally addressing that community. And of course, the hashtag as a form has spread to many other social and mobile media platforms—from Instagram and Facebook to text messaging, where it has developed many other uses, genres and meanings.

TOPICAL HASHTAG COMMUNITIES?

The extent to which any one group of participants in a hashtag may be described as a community in any real sense is a point of legitimate dispute. The term 'community', in our present context, would imply that hashtag participants share specific interests, are aware of, and are deliberately engaging with one another (which may not always be the case)—indeed, at their simplest, hashtags are merely a search-based mechanism for collating all tweets sharing a specific textual attribute, without any implication that individual messages are responding to one another (this is most evident in the case of emotive hashtags such as **#headdesk**).

On the other hand, there is ample evidence that in other cases, hashtags are used to bundle together tweets on a unified, common topic, and that the senders of these messages are directly engaging with one another, and/or with a shared text (or texts) outside of Twitter itself. Twitter users following and tweeting about recurring political events such as Question Time or the *Q&A* TV show in Australia, for example, about televised political debates in the U.S. presidential primaries, or simply about the stories covered by prime time news, *and using the appropriate hashtags as they do so*, are responding to shared media texts by using Twitter as an external backchannel for these broadcast media forms. Such users may not necessarily also follow what everyone else is saying about these same broadcasts, but they do take part in an active process of 'audiencing', as members of the community of interest for these shows (Highfield et al., 2013).

Twitter itself may also provide some more explicit evidence for community participation. It is possible, in particular, to measure the extent to which contributors to any given hashtag are actively responding to one another—by sending one another publicly visible @replies,[7] or retweeting each other's messages (in the case of manual retweets, possibly adding further commentary as they do so). A high volume of such response messages would indicate that users are not merely tweeting *into* the hashtag stream but also following what others are posting; the more such messages are contained in the hashtag stream—and the greater the total number of participants who engage in this way—the more the hashtag community can be said to act *as* a community. We suggest, in fact, that the ratio of responding to nonresponding hashtag posters may be especially valuable as an indicator of community: the fewer users who merely post into the hashtag without also responding to others, the more thoroughly connected the community might be.

Similarly, Bruns and Stieglitz (2012) have shown that hashtags relating to acute crisis events—from natural disasters through political unrest to other breaking news—exhibit a number of stable distinguishing properties that set them apart from other hashtags, for example, from those relating to major television events. Crisis hashtags are characterised by a high proportion both of tweets containing URLs and of retweets, pointing to a deliberate use of hashtagged tweets by users as a means of sharing emerging information about the crisis (in the form of URLs pointing to new updates) and more widely disseminating this information once found (by retweeting it). By contrast, hashtag activities around widely televised events—from popular shows through live sports to election night broadcasts—contain considerably fewer URLs or retweets, presumably because they are built around a shared primary text. The observation of these distinctly divergent, stable patterns in user activities for these different hashtag use cases points to the existence of different conceptualisations by users of the hashtag community that they are seeking to address or participate in.

It should also be noted that the hashtag community overlaps with other structural and communicative networks on Twitter. On the one hand, to regard tweets between two participants in a given hashtag conversation as evidence of a specific hashtag community may be to overestimate the importance of that hashtag, if the two users are also already connected as mutual followers—in such cases, the two users did not need to rely on the hashtag as a mechanism for discovering one another's tweets, because they also would have encountered them simply by reading their standard streams of incoming messages.

On the other hand, measuring the relative discursivity of hashtag conversations by identifying what percentage of hashtagged messages are responses to other users may also significantly underestimate the actual volume of user-to-user conversations which 'hang off' the hashtag: not all such responses will include the original hashtag, and lengthy conversations between two users who found each other through their shared use of a hashtag may follow on from that discovery but take place entirely outside of the hashtag stream itself. Even hashtags with comparatively low numbers of responses in the overall message stream may still engender significant levels of conversation between hashtag contributors, then—but outside the hashtag stream itself.

On balance, such over- and underestimations of the role hashtags play in enabling and stimulating conversations between participating Twitter users may cancel each other out; more importantly, what emerges from these observations is a picture of hashtag communities not as separate, sealed entities but as embedded and permeable macro-level spaces which overlap both with the meso-level flow of messages across longer-term follower/followee networks and with the micro-level communicative exchanges conducted as @replies between users who may or may not have found one another through the hashtag, as well as with other, related or rival hashtag communities at a similar macro level (cf. Bruns & Moe, 2014).

At a higher level of abstraction, the same may be said of Twitter and its variously defined communities of users (whether gathered around specific hashtags, closely interlinked as followers and followees, or connected through shared language, geography and other markers) and their position within the wider media ecology, alongside other social media platforms and alongside other forms of online and offline media. Twitter, too, is one among many interconnected social and traditional media platforms; it is neither entirely separate from them (since its constituency of users overlaps with theirs, and communication flows across their borders), nor completely homologous with them (since different sociotechnical affordances enable different forms and themes of communication).

The overall picture therefore resembles a 'network of issue publics' constituted via overlapping mediated public spheres (Bruns, 2008, p. 69). Those hashtags that emerge around a shared issue or interest should be seen as coordinating mechanisms for these issue publics—corresponding to, and in many cases also corresponding *with*, related issue publics as they may exist in other public spheres in areas such as politics, mainstream media, academia, popular culture and elsewhere.

HASHTAG COMMUNITIES AS AD HOC PUBLICS

What particularly allowed Twitter to stand out from such other spaces for issue publics from quite early on was the platform's ability to respond with great speed to emerging issues and acute events. In many other environments—especially those controlled by extensive top-down management structures—issue publics may form only post hoc: some time after the fact. Even online, news stories must be written, edited and published; commentary pages must be set up; potential participants must be invited to join the group. Twitter's user-generated system of hashtags condenses such processes to an instant, and its issue publics can indeed form virtually ad hoc, the moment they are needed. To include a hashtag in one's tweet is a performative statement: it brings the hashtag into being at the very moment that it is first articulated, and—as the tweet is instantly disseminated to all of the sender's followers (at least as long as the Twitter timeline is still organised chronologically rather than algorithmically)—announces its existence.

Not all hashtag communities constitute ad hoc publics, of course; some hashtag communities may even form praeter hoc, in anticipation of a foreseeable event (such as a scheduled television broadcast or an upcoming election), or come together only some considerable time after the event, as its full significance is revealed. However, it is this very flexibility of forming new hashtag communities as and when they are needed, without restriction, which arguably provides the foundation for Twitter's recognition as an important tool for the discussion of current events. This recognition is evident not least also in the utilisation of the platform by mainstream media organisations, politicians, industry and other 'official' interests, while the bottom-up nature of Twitter as a communicative space continues to be visible in the inability of such institutional participants to effectively channel or dominate the conversation.

The dynamic nature of conversations within hashtag communities provides fascinating insights into the inner workings of such ad hoc issue publics: it enables researchers to trace the various roles played by individual participants (for example, as information sources, community leaders, commenters, conversationalists, or lurkers), and to study how the community reacts to new stimuli (such as breaking news and new contributors). Such observations also offer perspectives on the interconnection of the community with other communicative spaces beyond Twitter itself and on the relative importance of such spaces; in all, they point to the overall shape of the event.

The specific dynamics of ad hoc communication in different hashtag communities diverge substantially, of course; different events and crises follow vastly different timelines, for example, and may attract considerably larger or smaller constituencies of participants, representing more or less diverse subsets of the overall Twitter user base.

Future research may engage with the gradual evolution of hashtag uses, especially also in relation to other mechanisms for coordinating topical discussions, both on Twitter and in other, competing spaces. For example, as follower networks on Twitter have developed and solidified, it is possible that many users likely to be interested in specific issues are already connected to each other as mutual followers and will encounter each other's topical tweets, increasingly rendering the explicit and visible use of hashtags as a means of flagging topicality unnecessary. But hashtags continue to be important as a mechanism for provoking the emergence of ad hoc publics around topics which are not well served by underlying network patterns (such as natural disasters, which are likely to affect users irrespective of their day-to-day topical interests), even if they may no longer be necessary for well-established topics of public debate.

AD HOC AND CALCULATED PUBLICS

Since their emergence in and adoption by Twitter as a platform, hashtags—their morphology, their cultural uses and their sociotechnical functions as interfaces to search engines and algorithms—have become incorporated into the cultural logics not only of Twitter but of the 'social media logics' (van Dijck & Poell, 2013) of the contemporary digital media environment more broadly. They are used within and across platforms (including Instagram, Tumblr and Facebook, as well as Twitter), and in each context their range of uses and meanings is slightly different. And within Twitter itself, the gradual changes to the platform's deployment of hashtags have transformed the ways they are experienced and the extent to which the publics that emerge and are coordinated via hashtags might be said to be truly 'ad hoc'.

For example, in 2012, Twitter began introducing official 'hashtag pages' for certain large-scale events, like NASCAR and the European Football Championships.[8] These pages (whose URLs appear just like any hashtag results page) are carefully curated representations of the public communication around an event, often privileging particular 'official' sources as part of Twitter's (then) new push to become more a media company than a social networking service. During Hurricane Sandy, for example, Twitter instituted a set of 'trusted' accounts whose tweets on the official hashtag #sandy would be privileged on the page returned following any search for '#sandy' (or after clicking on the hashtag in any tweet). On this official 'Sandy' page, Twitter provided lists of authoritative accounts to follow for information about the storm and offered free promoted tweets to key emergency response and recovery organisations.[9]

While the initial public communication around the gathering storm did 'self-organise' around the #sandy hashtag, thereby forming an ad hoc public as described above, Twitter, Inc., as well as various mainstream media, government and not-for-profit organisations added an additional layer of coordination and institutionalisation to the hashtag, not destroying or displacing the collective activity that constituted it but most definitely shaping and curating it. Of course, given the relatively open nature of Twitter as a communication platform, these attempts to make the Twitter stream around the emergency more useful and trustworthy can never really achieve the kinds of command-and-control structures that emergency services departments might prefer—witness the disruptive and playful activities such as fake or joke images that were still quite visible on the #sandy hashtag (Burgess, Vis, & Bruns, 2012).

While much of the early collective activity that gave the Twitter hashtag its meaning, including the meanings produced by 'ad hoc publics', can be understood as self-organising—especially in comparison to, say, Facebook and YouTube, where suggestive and social algorithms have long ordered our experiences of those platforms—the algorithmic turn is profoundly impacting on the affordances of the Twitter hashtag for public communication as well.

Twitter's search algorithm already profoundly shapes the results that users see. On most clients, a Twitter search will by default return a list of 'Top Tweets', which is some undisclosed cocktail of what the algorithm deems 'authoritative' or 'socially relevant' results. Some but not all clients also offer the user an 'All Tweets' alternative option. In the most systematic critique of social media algorithms so far, Tarleton Gillespie (2014) focuses particularly on fuzzy, value-laden concepts such as 'relevance'—'a fluid and loaded judgement, as open to interpretation as some of the evaluative terms media scholars have already unpacked, like 'newsworthy' or 'popular'" (p. 175)—and, we would argue, just as culturally powerful as those are. Gillespie points out that such evaluations made by algorithms 'always depend on inscribed assumptions about what matters, and how what matters can be identified' (p. 177). Here Gillespie notes the existing research that has identified 'structural tendencies toward what's already popular, toward English-language sites, and toward commercial information providers' in search engine results (p. 177), for example. Precisely these kinds of operations and underlying assumptions are in play when the Twitter platform displays a hashtag's search results.

Thus, while many users may click on a hashtag and assume that the resulting stream of tweets transparently represents the reality or even the totality of the tweets associated with a particular hashtag, they are (perhaps unknowingly) getting a constructed, partial and curated view of the tweets that have been posted as part of the conversation around that hashtag; and they are not privy to the basis on which this curation has taken place or how its associated choices have been made. It is only Twitter, Inc. that has what Gillespie evocatively calls 'backstage access' to 'the public algorithms that matter so much to the public circulation of knowledge' (p. 185). Indeed, it is hard to think of a better example of such algorithms than the ones that sort, order and curate the communicative material that constitutes the expression of issues or events around which publics emerge and engage. Whereas the first years of hashtags alerted us to the structural displacement of models like the 'public sphere' by terms like 'networked public' (boyd, 2010; Ito, 2008) and ad hoc publics, we now have to contend with the complications of personalised, *calculated* publics (Gillespie, 2014, p. 188) as well.

NOTES

1. This chapter is an updated and expanded version of Bruns, A., & Burgess, J. E. (2011). *The use of Twitter hashtags in the formation of ad hoc publics.* Paper presented at 6th European Consortium for Political Research General Conference, August 25–27, University of Iceland, Reykjavik. Retrieved from http://eprints.qut.edu.au/46515/. Project website: http://mappingonlinepublics. net
2. On this last, see also singh, Chapter 20, this volume.
3. The 11,800 used the #spill hashtag (see below); many more may have tweeted about the event without using the hashtag itself.
4. See Sauter and Bruns, Chapter 3, this volume.
5. On the use of meta-hashtags and tag co-occurrence, see Rocheleau and Millette, Chapter 18, this volume.
6. See Peck, Chapter 13, this volume, and Rambukkana, Chapter 2, this volume, for discussions of these two tags respectively, including how even humorous tag conventions can be drawn into political publics (e.g., #RaceFail).
7. @replies are tweets which contain the username of the message recipient, prefixed by the '@' symbol. Private, direct messages would also indicate engagement between community members, of course, but such messages are not commonly available to researchers, for obvious reasons.
8. See for example https://blog.twitter.com/2012/euro-2012-follow-all-the-action-on-the-pitch-and-in-the-stands
9. https://blog.twitter.com/2012/hurricane-sandy-resources-on-twitter

REFERENCES

boyd, d. (2010). Social network sites as networked publics: Affordances, dynamics, and implications. In Z. Papacharissi (Ed.), *Networked self: Identity, community, and culture on social network sites* (pp. 39–58). New York: Routledge.

Bruns, A. (2008). Life beyond the public sphere: Towards a networked model for political deliberation. *Information Polity, 13*(1–2), 65–79.

Bruns, A. (2010a, December 30). Visualising Twitter dynamics in Gephi, part 1. Retrieved from http://mappingonlinepublics.net/2010/12/30/visualising-twitter-dynamics-in-gephi-part-1/

Bruns, A. (2010b, December 30). Visualising Twitter dynamics in Gephi, part 2. Retrieved from http://mappingonlinepublics.net/2010/12/30/visualising-twitter-dynamics-in-gephi-part-2/

Bruns, A., & Moe, H. (2014). Structural layers of communication on Twitter. In K. Weller, A. Bruns, J. Burgess, M. Mahrt, & C. Puschann (Eds.), *Twitter and society* (pp. 15–28). New York: Peter Lang.

Bruns, A., & Stieglitz, S. (2012). Quantitative approaches to comparing communication patterns on Twitter. *Journal of Technology in Human Services, 30*(3–4), 160–185. doi:10.1080/15228835.2012.744249

Burgess, J. (2010). *Remembering Black Saturday: The role of personal communicative ecologies in the mediation of the 2009 Australian bushfires.* Paper presented at ACS Crossroads, June 17–21, Lingnan University, Hong Kong.

Burgess, J. (2012). The iPhone moment, the Apple brand and the creative consumer: From 'hackability and usability' to cultural generativity. In L. Hjorth, I. Richardson, & J. Burgess (Eds.), *Studying mobile media: Cultural technologies, mobile communication, and the iPhone* (pp. 28–42). London: Routledge.

Burgess, J. (2014). From 'broadcast yourself' to 'follow your interests': Making over social media. *International Journal of Cultural Studies* [online first]. Retrieved from http://ics.sagepub.com/content/early/2014/01/13/1367877913513684.abstract

Burgess, J., & Crawford, K. (2011). *Social media and the theory of the acute event.* Paper presented at Internet Research 12.0—Performance and Participation, October 9–11, Seattle.

Burgess, J., Vis, F., & Bruns, A. (2012, November 6). How many fake Sandy pictures were really shared on social media? *Guardian Datablog.* Retrieved from http://www.guardian.co.uk/news/datablog/2012/nov/06/fake-sandy-pictures-social-media

Deller, R. (2011). Twittering on: Audience research and participation using Twitter. *Participations, 8*(1). Retrieved from http://www.participations.org/Volume%208/Issue%201/deller.htm

Gillespie, T. (2014). The relevance of algorithms. In T. Gillespie, P. Boczkowski, & K. Foot (Eds.), *Media technologies: Essays on communication, materiality, and society* (pp. 167–194). Cambridge, MA: MIT Press.

Highfield, T., Harrington, S., & Bruns, A. (2013). Twitter as a technology for audiencing and fandom: The #eurovision phenomenon. *Information, Communication & Society, 16*(3), 315–339.

Huang, J., Thornton, K. M., & Efthimiadis, E. N. (2010). Conversational tagging in Twitter. In *HT '10: Proceedings of the 21st ACM Conference on Hypertext and Hypermedia* (pp. 173–178). New York: ACM. Retrieved from http://portal.acm.org/citation.cfm?id=1810647

Hughes, A., & Palen, L. (2009). Twitter adoption and use in mass convergence and emergency events. *International Journal of Emergency Management, 6*(3–4), 248–260.

Ito, M. (2008). Introduction. In K. Varnelis (Ed.), *Networked publics* (pp. 1–14). Cambridge, MA: MIT Press.

Marres, N. (2012). *Material participation: Technology, the environment and everyday publics.* London: Palgrave.

Messina, C. (2007a, August 25). Groups for Twitter; or a proposal for Twitter tag channels. *FactoryCity* [Blog]. Retrieved from http://factoryjoe.com/blog/2007/08/25/groups-for-twitter-or-a-proposal-for-twitter-tag-channels/

Messina, C. (2007b, October 22). Twitter hashtags for emergency coordination and disaster relief. *FactoryCity* [Blog]. Retrieved from http://factoryjoe.com/blog/2007/10/22/twitter-hashtags-for-emergency-coordination-and-disaster-relief/

Rogers, R. (2013). Debanalising Twitter: The transformation of an object of study. In K. Weller, A. Bruns, J. Burgess, M. Mahrt, & C. Puschmann (Eds.), *Twitter and society* (pp. ix–xxvi). New York: Peter Lang.

Schmidt, J. (2014). Twitter and the rise of personal publics. In K. Weller, A. Bruns, J. Burgess, M. Mahrt, & C. Puschmann (Eds.), *Twitter and society* (pp. 3–14). New York: Peter Lang.

Small, T. (2011). What the hashtag? A content analysis of Canadian politics on Twitter. *Information, Communication & Society, 14*(6), 872–895.

Van Dijck, J., & Poell, T. (2013). Understanding social media logic. *Media and Communication, 1*(1), 2–14.

Warner, M. (2005). *Publics and counterpublics.* New York: Zone Books.

CHAPTER TWO

From #RaceFail to #Ferguson: The Digital Intimacies of Race-Activist Hashtag Publics[1]

NATHAN RAMBUKKANA

One can argue the merits of Habermas's public sphere—that is, one where rational critical discourse on matters of societal importance (such as, most critically, the actions of the state) can take place; populated by citizens stepping out of their private roles as interested individuals and into a public space where they become participants in disinterested discussion and debate (Habermas, 1962/1989). But what of the other kinds of discussion and debate that are facilitated by networked technology? Taking its cue from critical public sphere theorists such as Nancy Fraser (1992) and Michael Warner (2002), this paper explores those *other* publics: more or less subaltern, more or less rational, more or less critical, and almost certainly partial, affective, interested, and *loud*. It's interested in angry publics. It's interested in fringe publics. It's interested in the kinds of publics that do politics in a way that is rough and emergent, flawed and messy, and ones in which new forms of collective power are being forged on the fly, and in the shadow of loftier mainstream spheres. These are the publics born of frictions, in Anna Tsing's sense, "the awkward, unequal, unstable and creative qualities of interconnection across difference" (2005, p. 4).

Specifically, this paper unpacks the publics of the hashtag, that piece of multiply repurposed typography, that rebel punctuation mark moving to establish itself in new regimes of discourse and communication. Like other forms of active punctuation such as the @, $, and !, those residents of QWERTY keyboards that, out of sheer convenience, were deployed by programmers and users alike to take on additional, performative meanings, the hashtag is now one of the most recognizable symbols of communication itself. From its humble Roman origins, where it was used to denote that something was measured in pounds—a meaning that drifted towards also meaning something numbered (Roman, 2014)[2]—it was twice appropriated by early digital communication cultures. The first appropriation was in 1969, as the "redial" button on early touch-tone phone keypads, where it was known as the "octothorpe" (Roman, 2014), and the second was in 1988, as a symbol marking out IRC channels (Zappavigna, 2011, p. 791). From its digital neonacy, it was then mobilized by Chris Messina, who, inspired by folksonomic endeavours such as del.icio.us, selected it due to its ubiquity on pre-smartphone keypads (pers. comm.) and proposed its use as a ground-up search symbol in tweets, a structure that was later worked into Twitter's interface as an official function (Messina, 2007; Roman, 2014). Now, moving beyond Twitter alone and into other structures both on and offline, it has the potential to organize new structures of discussion, new "ambient affiliations" (Zappavigna, 2011, p. 803), and new "potential discourse communities" (p. 801) where in-line metadata draws discourse together *across technologies.*

Hashtags are hybrids in the taxonomy of types of information. They are both text and metatext, information and tag, pragmatic and metapragmatic speech. They are deictic, indexical—yet what they point to is themselves, their own dual role in ongoing discourse. The hashtag functions in the space between the contextual and the chronological (Benovitz, 2010, p. 124). It's a node of continued context across media, conversations, and locales, and yet it emerges temporally, self-developing through time, pointing to itself as it points to the other texts it marks as within its ambit.

In this way, hashtags push the boundaries of specific discourses. They expand the space of discourse along the lines that they simultaneously name and mark out. New hashtags are constantly being deployed: some may not "catch" and will simply sink back into the webwork; others might rise to prominence through repetition, through use, through uptake. We could contrast this form of data organization to processes such as agenda setting, gatekeeping, and creating news frames where those controlling messages set the shape of the discursive content from above, limiting the ways in which information flows may be organized and presented (Bastos, Raimundo, & Travitzki, 2013, p. 264). While not discursive limiters or boundaries, then, hashtags do have the ability to mark the discursive flows of an event and are in fact events themselves (Rambukkana, Introduction, this volume; Sauter & Bruns, Chapter 3, this volume), straddling that dual role as text and metatext.

Drawing on thinkers such as Massumi (2011), Kwinter (2001), and Whitehead (1967), events are moments when otherwise disparate experiences entwine. Threads of as yet unrelated material flow together in complex aggregates and assemblages that, once together, take on a novel character. They move from the complex to the singular, twining into threads of their own that go on to combine into further complexes, further assemblages. The gift that Foucault (1972) gave us for the figuring of discourse and discursive power was to draw our attention to discourse as *event*. As researchers, to understand the power and embeddedness of a specific discourse, we need to trace it, to discover both its contextual relevance and its temporal arc—to trace *both* its affects *and* its points of origin, points of change. Where event theory can add to our understanding of the discursive is in showing how, in our current cross-linked and metatextual environment, events of discourse are coming to *recognize and mark themselves as such*. This self-reflexive discourse occurs in multiple ways—not the least of which is the popular uptake of the term "discourse"—but one worth highlighting due to its rising societal prominence is the political and discursive use of the hashtag.

Previous studies have looked at the role of hashtags in relation to political discourse. Most of these are centred around analysing how the hashtag operates on Twitter during major—but temporally limited—political events. Hashtags such as **#Jan25, #Tahrir, #Egypt, #Tunisia,** and **#spill** have been analysed, for example, with respect to their role in the Arab Uprisings (e.g., Papacharissi & Oliveira, 2012) and in a 2010 leadership upset in the Australian parliament (Bruns & Burgess, 2012). On a short timescale, events and studies such as these demonstrate how the hashtag can work as a uniting thread of discourse that allows those who use it to feed into an ongoing and evolving conversation—even, as Douai (2013) points out, acting as one of the seeds that can contribute to change. This paper, however, looks at the political hashtag on longer timescales, and for events that are ongoing. It looks at places where tensions simmer and build up over time, where frictions develop and ignite, and where those tracking these fraught politics find new problematics and issues, drawing attention to them to wear away at normativities and hegemonies. It also looks at how this kind of hashtag can spill beyond Twitter into other spaces of social networking—such as the recent uptake of hashtag mechanics on Facebook and its more established uses on Instagram and Tumblr—and even to other media spaces such as television and print, billboards, skywriting, and graffiti. But how exactly can we situate the hashtag with respect to political communication?

COMMUNICATION, FRICTION, AND DIGITAL INTIMACIES

Communication is not a process that naturally invites understanding, connection, or societal harmony—despite a common conception that this is the case (Peters, 1999). Digital convergence increases the sphere of global connection, but that does not, in and of itself, heal a world community broken by misunderstanding.

Rather, as Peters underlines, divergent understandings and interests are the natural baseline. The political role of communication media cannot, therefore, be to "fix" a broken system of public-sphere communication, but rather to "unfix" staid communication patterns, to refigure the public conversation about important issues and topics (such as inequality, racism, sexism, and abuses of power) with a view to cracking open stable systems of meaning making, and working—as Peters reminds us we need to do (1999, p. 30)—to *build* better communication across and between cultural and subcultural spaces. We need to work with and through the "productive friction[s] of global encounters" (Tsing, 2005, p. 3), to tap the unpredictable and raw energy that is concomitant with our increasing interconnection.

And an increasingly prominent venue of such connections is in the realm of "digital intimacies." Intimacies, in the critical sense, are the "kinds of connection that *impact* on people, and on which they depend for living" (Berlant, 1998, p. 284; emphasis in original). For Lauren Berlant, the space of intimacy comprises close connections that matter and that subtend our lives and experiences, define multiple forms of human relationship, and act as crucial spaces of mediation between our selves and our worlds. The realm of the digital affords a multiplicity of new forms of intimate connection, especially now since, as Vincent Mosco discusses, digital technologies have sunk "into the woodwork" (2004, p. 21) and become part of the texture of our everyday lives. Nancy Baym even goes so far as to call cyberspace a myth (2010, p. 150); from her perspective, there is no separate space of digital connection, not any more. The digital has simply become one of the regular conduits of connection for everyday life—digital intimacies are now, quite simply, part of the new shape of human intimacy.

And the intimacies of our lives *affect* us; they forge, sustain, renew, break, or alter connections that matter in our lives. And this is doubly so for intimate linkages and disassociations that might be working at an unconscious level, that tap their affective vigor from some of the more problematic elements of status quo relationships, such as from racist, sexist, classist, and ableist "common sense" understandings. This is why it is important to investigate not only intimacy, but also "intimate privilege"[3] (Rambukkana, 2015). One of the challenges in thinking through digital intimacies, then, is to think about what forms of intimacy might be privileged in the digital realm, as well as how the intimate privileges of society, and resistance to them, surface in aspects of digital culture.

This brings us back to the hashtag, that technosocial event (Rambukkana, Introduction, this volume) that taps one of the privileged aspects of digital culture, virality, and uses it to promote and extend everything from advertising campaigns to activist engagements. In a world of increasing and cacophonous complexity, hashtags are singular moments that coalesce into something new: threads of meaning that work to weave new abstractions into the world. These new elements can range from the ridiculous to the sublime, from the politically irrelevant to the politically potent. It is this deeper end of the meme pool that is worth investigating further: What are these broader political functions of the hashtag? And how might something like race-activist hashtags form a kind of "sticky engagement" (Tsing, 2005, p. 6) with these issues through time, and over distance? We will explore these questions through a singular but ongoing discursive event, RaceFail, and its affinities with subsequent tags that call out and expose racism in multiple venues of life and culture, most notably, **#Ferguson**.

THE GREAT CULTURAL APPROPRIATION DEBATE
AND RACEFAIL OF AUGHT NINE

We can consider as a case study that demonstrates the power of the hashtag as political discourse the sustained conversation about RaceFail that started in the science fiction and fantasy (SFF) blogosphere. Methodologically, this exploration mobilizes Rogers's suggestion to "follow the medium" as a way to "reorient Internet research to consider the Internet as a source of data, method and technique" (qtd. in Bruns, 2012, p. 1329). By taking the object of study, "**#RaceFail**/ RaceFail" (with and without its "hashed" diacritic), and employing multimodal searches for its incidence as its major data-gathering methodology, this paper uses what might be termed a snowball textual analysis sample to gather material across media linked through the same tag. Unlike many early studies of hashtags that use programs such as TwapperKeeper to track and record tweets (e.g., see Bruns, 2012, p. 1331), this study centres the life of this tag in ways that contain but exceed the scope of the Twitterverse. The following texts were gathered using a number of tools, including Topsy, a website that provides the ability to search across social media platforms and offers useful advanced search tools that, for example, allow searching the timeline for the earliest posts; as well as Google, for its far-reaching search tendrils and metrics; and HootSuite in combination with Google Spreadsheets and TAGS Explorer to track and visualize more recent incidents on Twitter. Together, and through the cross-linking and archive posts found in many of the major blog posts on this topic, we can piece together a picture of this discursive event as it first exploded and then developed into touchstones—the **#RaceFail** hashtag and the word RaceFail itself—that continue to be appended to ongoing discussions, working to evoke the spirit of the earlier debate and simultaneously stitch in new material.

"The Great Cultural Appropriation Debate and RaceFail of Aught Nine" stretches back as a named discourse to January 2009. It was (and to a certain extent, still is) a messy, involved, heated, emotive discussion. In other words, it's the complete opposite of what Habermas would call "rational critical discourse"

(1981/1984), and would be seen through his lens as a misuse of public-sphere resources. But for several subaltern counterpublics (such as the SFF blogosphere, the antiracist blogosphere, and fan publics of multiple stripes) RaceFail '09 was a seismic shift in subcultural norms, trends, and discussion. It kicked off a more visible critical race mediascape in SFF writing and criticism, as well as on the net in general.

This online quarrel started in reaction to a blog post by speculative fiction author Elizabeth Bear about the politics and process of "writing the Other" in science fiction and fantasy. Her post downplayed the fraught politics of representation and presented a facile list of pitfalls to avoid and guidelines to follow to access what she characterises as a "simple, but not easy" process (Bear, 2009). Her dismissal of the more nuanced arguments about the power relations and history of representing People of Colour (PoC) and Indigenous people, coupled with what some read as racist narrative elements in her own work (Avalon's Willow, 2009), sparked first negative comments, then response posts containing further negative criticism. Collectively, these discussed (often heatedly) issues such as racism, cultural appropriation, colonialism, Orientalism, exoticism, and White privilege—both in relation to SFF texts themselves and the norms of the SFF publishing industry—after which things got ugly (Somerville, n.d.). Those critiquing Bear were themselves virulently critiqued, mostly by what seem to be White fans/authors/publishing executives. The main tension (the major "Fail") was how writers, fans, and critics of colour were attacked about having the right to critique at all and characterised as lacking the required intelligence and decorum for "proper" intellectual debate. After this, the quarrel descended into further personal attacks, such as calling the PoC critics and allies "trolls" or "orcs" and threatening to "blacklist" them from industry publishing and fan culture (Somerville, n.d.).

Collected around these loose main structuring events in the blogosphere is a nimbus of related discourse. The Topsy timeline indicates that while there were discussions of race in science fiction and fantasy before this epoch-making epic web brawl, the emotions and issues thrown up in this scuffle were truly an "incitement to discourse" in Foucault's (1976/1990, p. 17) sense, with the keyword and then hashtag RaceFail—not present in web archives before this point—crystalizing the affect of these debates into a diachronic anchor point that allows the thread to be pulled together across space and time. It's interesting, in particular, how this tagging had the ability to pull even-earlier material into its ambit, such as Bear's original post. By subsequent diachronic linking and archiving and synchronic blog posts and comment threads, the hashtag collects and self-organizes the material most relevant to it, linking back to incorporate even material that came before it and helped it become.

It's this linking function that is perhaps the most interesting. Now you can find the #RaceFail hashtag across media platforms, and branching out to new subject matter. A review of Topsy findings collects material from Twitter, personal blogs, news blogs, mainstream news sites, aggregators such as Tumblr and Reddit, and some wikis. At a second level of analysis, these posts and tweets, especially the later ones, utilize **#RaceFail** to signify a discussion of racism in other spheres. It is in this sense that race-activist tagging, in addition to branching off to incorporate multiple media, starts to develop beyond the SFF public sphere to address other small publics as well as the mainstream public sphere as a whole. For example, there is discussion of mainstream movies, television, and news, both within the realm of sci-fi and fantasy and beyond, with topics such as: the whitening of casting in *The Last Airbender* (Shyamalan, 2010) and *The Hunger Games* (Ross, 2012); the cultural appropriation of various popular Halloween costume tropes; racist comments by or about politicians, actors, or sports figures; and even how race crosses with other issues, such as the politics of the SlutWalk. Recent clusters show it being used, for example, to discuss the racist slave liberation narrative on HBO's *Game of Thrones* (Benioff & Weiss, 2011), an unfortunate yellowface incident on CBS's *How I Met Your Mother* (Bays & Thomas, 2005)—that was also trending as a topic through the hashtag **#HowIMetYourRacism** (Digital Spy, 2014)—and the whitewashing of Aronofsky's biblical epic *Noah* (2014; Smietana, 2014). Sometimes the hashtag is in the text itself, sometimes in the comments it incites, but every time the tag is used—and Google Trends show over 108,000 individual hits from a peak interest in mid- to late 2009 through to mid-2014—a piece is added to the discussion.

However, this isn't to say that this discussion is unified, consistent, or even entirely on topic. In fact, despite its persistence, the channel is a muddy one. For example, while the predominant use of **#RaceFail** is as a race-activist hashtag, an important secondary usage rubbing up against it is about "Fail" moments in relation to running culture, such as "Aaaaaand it's raining **#racefail**" (@SCRunnerGrl, 2015). Another parallel usage is a recent cluster that saw **#RaceFail** being taken up to discuss a "Fail" incident on the program *The Amazing Race* (Doganieri & van Munster, 2001), leading **#AmazingRace** to be the top hashtag used alongside **#RaceFail** from September 2013 to March 2015,[4] an aberration that caused friction among the multiple publics it networked together, such as one user who tweeted "**#RaceFail** usually means something very different in my circles **#AmazingRace**" (@haymakerhattie, 2014). Such usages, while they do "muddy the waters," are not deliberate attempts to swamp the discussion and are perhaps a low-level frictive force we could think of as a form of noise. Another fraught element is how, while **#RaceFail** does have a consistent presence across multiple platforms, it also often appears as RaceFail, without the symbol, meaning that parts articulated to this hashtag public will only incidentally and inconsistently be collected by following the tag (for example, when an article is linked to or a tweet retweeted). The fact that this hashtag public is not even always hashtagged as such—even on Twitter—is a further friction that underlines its provisional nature as a consistent public.

Yet despite the inconsistencies and the misdirections, this discursive channel has still carved out a distinct and persistent political voice—but what does that critical voice *do*? How does this welling up of critical race discourse function with respect to those who deploy it, for fandom, for science fiction and fantasy, and for society in general? As N. K. Jemisin points out, it contributed to paradigm shift in these discursive spaces. On the rough anniversary of the explosion of RaceFail, critical race blogger and acclaimed fantasy writer N. K. Jemisin published a blog post titled "Why I Think RaceFail Was the Bestest Thing Evar for SFF" (2010). In it she talks about how, contrary to some who think the "Fail" part of RaceFail was that it happened at all, these issues needed to be discussed. But more: they needed to be *argued*. As Sara Ahmed notes, the policing of affect can cause negatives from psychological distress to bad politics. Subjects in precarious circumstances, such as immigrants, refugees, sessional workers, are often barred from expressing legitimate anger and are even forced to perform a kind of "coercive happiness" in order to receive or retain access to common resources (2009). Fan cultures—and SFF fandom is particularly known for this—can perform some of that same policing, where anger at subcultural privilege and oppression around issues such as sexism and racism traditionally has been met with derision and exclusion.

Take, for example, **#GamerGate**. **#GamerGate** began in 2014 as a targeted misogynist attack against game developer Zoe Quinn, involving public discussions of her sex life and accusations of collusion with game journalists to promote her game *Depression Quest* (Hathaway, 2014). This backlash against feminist games and game criticism became fused with the earlier backlash against Anita Sarkeesian's (2012) Kickstarter campaign to start a web series about sexism in video games, to which the response from some fans was the creation of sexist and misogynist memes, multiple death and rape threats, and even a flash video game the point of which was to beat up Sarkeesian to "put her in her place" (Chee & Bergstrom, 2013). This backlash, which has been framed by some participants as a symposium on ethics in games journalism (Hathaway, 2014), has become an all-out brawl, with much anger on both sides. With this kind of staid culture, anger and intense discursive fighting are perhaps the only ways to bring about true change. As with **#GamerGate** and the problem of sexism in video game culture, the quarrel—from the Latin *querel* (complaint)—that underlay the question of race representation in SFF was always there waiting to happen; it was a disconnect, an *aporia*, a problematic issue that needed a prolonged and angry argument to air its grievances and map out their discursive extent. And as a discursive event, it bore fruit.

Jemisin notes that since RaceFail, there has been an "increased awareness [in] the SFF zeitgeist on race issues" (2010) that has led to many positive outcomes, including some publishers actively seeking out authors of colour (risking token-ism, but also addressing a long-identified structural bias against authors of colour in SFF). It has also led to generally more attention being paid to issues of race in SFF, such as Nalo Hopkinson giving the keynote address at the International Association for the Fantastic in the Arts conference in 2010 and Jemisin being Guest of Honour at Continuum in 2013. Jemisin's point is that these changes would not have occurred in the absence of the uppity spirit of #RaceFail. From her perspective, RaceFail worked as a solidarity movement that "turned up the heat" on an underaddressed issue (2010). It shook up a few stable systems (such as those of SFF fandom and publishing), creating new opportunities for ongoing discussion and debate that are having an ongoing material impact on the shape and nature of those publics.

More recently, during her 2013 speech as Guest of Honour at Continuum in Australia, Jemisin called for a discourse of reconciliation to keep alive this powerful discussion and, tapping the Australian model of reconciliation, to try to redirect it into something positive, inclusive, even healing (2013). She contrasts Australian attempts at reconciliation with the situation in the U.S., where some are aggressively trying to repeal Black voting rights, where Stand Your Ground laws and similar in places such as Florida and Texas "allow White people to shoot Black people with impunity," as in the cases of Trayvon Martin and (we could add) Michael Brown, and where even a fraught and limited reconciliation process would be an impossibly distant pipe dream (Jemisin, 2013). While the metaphor of Australia as a troubled political space on the road to rediscovery and renewal is perhaps a bit too simplistic, where she takes this argument with respect to SFF is more fruitful. Her call for reconciliation within these genres, at the levels of both fandom and publishing, is a call to harness the unease and anger into something generative, to use it to reclaim supressed but central threads—to reweave the tex-ture of SFF discourse into a stronger, more enduring fabric.

And among the elements she draws on in discussing how this could happen is the event of RaceFail in its ongoing digital persistence. She writes: "If you did not follow RaceFail when it occurred or if you dismissed it as too much to handle, try. It's all still there; just Google it. Hundreds of people poured millions of words into articulating what's wrong with this genre, and how those wrongs can be made right. You owe it to yourself to read some of what they wrote" (Jemisin, 2013). Her words, both in the speech and in its archival on her blog and various other spaces, further link this discussion to the tag of RaceFail, adding a further articulation to its ongoing meaning.

As to how such discursive changes and resistances surface in broader publics, tracing some of the more recent uses of #RaceFail brings us examples such as the apology for the Yellowface incident on *How I Met Your Mother*. Akin to the Habermassian ideal of how the public sphere operates with respect to state power, the publicity that accrues to RaceFail-articulated discussions is having some effect in the broader spheres of cultural production and political discussion. For example, following the Yellowface incident, a flurry of tweets as well as posts in multiple other media (e.g., Angry Asian Man, 2014; Racialicious, 2014) led showrunners Carter Bays and Craig Thomas to issue the following statement through a series of tweets:

> Hey guys, sorry this took so long. @himymcraig and I want to say a few words about **#HowIMetYourRacism**
>
> With Monday's episode, we set out to make a silly and unabashedly immature homage to Kung Fu movies, a genre we've always loved.
>
> But along the way we offended people. We're deeply sorry, and we're grateful to everyone who spoke up to make us aware of it.
>
> We try to make a show that's universal, that anyone can watch and enjoy. We fell short of that this week, and feel terrible about it. (Digital Spy, 2014)

While the motivations behind this statement may be genuine or rhetorical, the effect of such an acknowledgement could bode well for more mindful future cultural production—if only to avoid bad press.

In the era of RaceFail, such decisions are now increasingly being challenged and talked about; they no longer happen in a cultural vacuum where privileged representations of difference, race, and world cultures simply become *the* representations. As Jeff Rosenblum argues in the film *The Naked Brand* (Huang & Rosenblum, 2013) with respect to the impact of social media criticism on corporations, in this era of high networked visibility through new media, public scrutiny can act as a powerful motivator of corporate action—a way maybe to harness some of the pervasive energy of neoliberalism and wrest it towards effecting (perhaps reluctant) social justice and social change. And, as with corporations, for the creatives working to fashion fictions, these public frictive reflections on representation will continue to have an impact on the stories they are putting out into the world. This matters because the stories we tell ourselves, even the science fictional and fantastic ones set in other worlds, are still populated with images and tropes of race and difference from the world we actually live in.

#RaceFail acts as a persistent marker for these ongoing public-sphere discussions. While the **#Fail** hashtag predated RaceFail and was considered by some a persistent "'microgenre' [used for] complaining about something, often technological" (Zappavigna, 2011, p. 803), **#RaceFail** couples this persistent genre with a persistent politics, one that allows those ongoing discussions to weave their complexity into a singular—and powerful—discursive event. This kind of "sticky engagement" (Tsing, 2005, p. 6) can even spill over into cognate spheres of discourse. We now turn to a brief consideration of one such parallel discussion, that around **#Ferguson**.

#FERGUSON AND THE PERSISTENCE OF RACE-ACTIVIST HASHTAG PUBLICS

When police officer Darren Wilson shot and killed unarmed teenager Michael Brown on 9 August, 2014, in the small town of Ferguson, Missouri, the response of the predominantly African American St. Louis suburb was one of disbelief, anger, and protest (Kroh, 2014). While the story as a whole has multiple facets, extended over months and involving protests, a disproportionate police response, and synergistic protests catalyzed around subsequent events, such as the death of another Black teen, Vonderrit Myers Jr., in October (Covert, 2014), the contours are all too familiar. The anger that emerges from these situations is visceral and vital, forged in long histories of injustice, resistance, and exhaustion. Such is the raw amplitude of this frustration and anger that it imbues the social media discussions that reflect it with these qualities, as the affect of these struggles pours into this new medium of public discussion. Just as RaceFail exploded into the relatively new social spaces of the blogosphere, the shooting in Ferguson and its continuing aftermath have found a powerful articulation through social media discussions and in particular through the mobilization of **#Ferguson**.

#Ferguson is the ninth most frequently used hashtag in conjunction with **#RaceFail** from September 2013–March 2015, and arguably the most resonant one in mainstream news media, both U.S. and international. Deployed soon after news broke about the events, the tag grew to over 4,000 hits in only 9 hours (Zerehi, 2014). Nor is it the only race-activist hashtag to emerge from these events. Others, such as **#FergusonOctober, #IAmMichaelBrown,** and **#BlackLivesMatter,** to mark the large October protest and highlight the repetitive occurrence of White police officers killing Black teens in the U.S., and **#ArrestDarrenWilson** and **#JusticeForMikeBrown,** to call for justice in this case, were all prominently featured as part of these protests—and not just within tweets but also on other social media platforms and even in analogue media such as protest signs (see Figure 2.1).

The space of the racialized hashtag public is one of powerful and mobile affectivity, drawing together and separating, calling out and memorializing, injecting a biopolitical problematic into a new sphere of human interaction. The Ferguson shooting and protests were featured on international and transnational media (e.g., Al Jazeera, 2014; Petersen, 2014); one BBC news article noted that Amnesty International was sending observers to monitor police actions and support the communities (a first for the U.S.), and that critics of the U.S. were piling on, with Iran's Supreme Leader Ali Khamenei tweeting criticism of how the U.S. was treating its own citizens, and using **#Ferguson** in the tweet (Petersen, 2014). As Sauter and Bruns (Chapter 3, this volume) note, "Thematic hashtags rarely exist in isolation from a wider discursive context: they emerge out of, are part of, and can shape events in the wider online and offline world." While the social media articulation of #Ferguson did not "cause" this struggle to become global news, it is a significant part of its global articulation. Like how the portable video camera became a crucial part of the event that was the Rodney King police beatings and their fraught aftermath, the political use of hashtags helped ignite a global reflection on race, the policing of Black bodies, and the militarization of police and became a crucial part of this new assemblage—despite and in direct competition to the neoliberal tendency of trying to transform the hashtag into an appendage of the advertising industry.

Figure 2.1. France François participates in the DC Rally for Mike Brown (Image used by permission of subject). (Starr, 2014).

The tension between this neoliberal desire to reduce hashtag publics to product publicity and the activist desire to use hashtags to further public-sphere awareness of political issues is codified in the controversy over the "algorithmic filtering" (Tufekci, 2014) of #Ferguson on Facebook. When many began to notice that #Ferguson was trending on Twitter but absent or near-absent on Facebook news feeds, critics started to reflect on the reason. One of the most prominent critics, Zeynep Tufekci, asks:

> But I wonder: what if Ferguson had started to bubble, but there was no Twitter to catch on nationally? Would it ever make it through the algorithmic filtering on Facebook? Maybe, but with no transparency to the decisions, I cannot be sure.
> Would Ferguson be buried in algorithmic censorship? (Tufekci, 2014)

The question of Facebook's algorithmic filtering and its effect (willed or not) spilled into mainstream media, with one CBC News article citing Anatoliy Gruzd, who noted that "there's a concern that the algorithmic filtering of the content can hinder the ability of the protesters to build awareness for their particular campaign" through the use of hashtags on Twitter and Facebook (qtd. in Zerehi, 2014). The same article cites Fenwick McKelvey, who notes that this is not a new debate, with Twitter as well garnering criticism for its opaque and proprietary filtering practices. He notes that it might not be "editorial bias that dictates what appears or doesn't appear on our timeline[s, but rather] algorithms intended to help us get a personalized experience on the web" (qtd. in Zerehi, 2014). The collective point of these articles and voices is that such filtering practices affect and shape our personal mediascapes—including, crucially, whether or not we will see controversial, fractious, or political content—and as such, our abilities to engage with such issues. While one can argue that it might not be incumbent on sites such as Facebook to curate our entire news consumption, the fact that software algorithms are being used to direct us toward certain content and away from others—or even to manipulate our moods (Meyer, 2014)—raises concern for issues as broadly situated as net neutrality, public-sphere access, research ethics, and digital media futures.

The social media mobilization around #Ferguson has led to an unprecedented response, including international news coverage and even transnational policy intervention, such as on 11 November, 2014, when Michael Brown's parents spoke in front of the UN Committee against Torture (Townes, 2014). This ongoing story would likely not have had the same sustained impact and international reverberation without its social media amplification—one that no doubt had a catalyzing effect for those on the ground as well, acting as a global echo chamber for the sustained event of this loud and angry protest. At time of writing, people were waiting for the grand jury decision in the case of Darren Wilson,[5] and Missouri governor Jay Nixon was poised to call in the National Guard and also noted that over 1,000 police officers had completed over 5,000 hours of specialized training to prepare for this verdict, leading some commentators to note that given the preparation, the verdict seemed to be a foregone conclusion (Townes, 2014). Nixon went on to announce that (ironically) these police measures were put in place so that peaceful protestors would not have their rights violated, adding that "violence will not be tolerated" (qtd. in Fairbanks, 2014). This statement, so out of step with Missouri's violent response to the first, 90-day phase of this protest, and with the violent nature of policing of People of Colour in general, prompted journalist Julia Carrie Wong to post a series of images depicting victims of police violence with the repeated legend "violence will not be tolerated" (Fairbanks, 2014), one soon taken up as new viral hashtag often articulated in combination with #Ferguson:

> #ViolenceWillNotBeTolerated @GovJayNixon Police fire teargas and rubber bullets into peaceful #Ferguson protest. http://youtu.be/GO1SKC6dK7o (Fairbanks, 2014)

And when the Ku Klux Klan (KKK) reared their unwelcome cowled heads as part of this discussion, announcing that they would use lethal force against Ferguson "terrorists," Anonymous launched "Operation KKK" to take over or block KKK social media, as well as to unhood and expose its prominent members (Beaumont, 2014). Although still not a complete event, it is clear that this is a messy fight, and one in which social media in general, and race-activist hashtags in particular, are playing crucial parts.

CONCLUSION: BEYOND MERE SEARCHING

Searching and tagging are powerful technologies. They allow us to cut across contexts and gather material, texts, resources. But beyond the individual process of searching and gathering like digital data, it is the self-consciousness and prominence of the hashtag that marks it as something worth noting—both in general and with respect to its potential impacts for activist politics and organizing. While folksonomic tagging ventures and attempts to organize and archive conversations, debates, and fields of thought all break open exciting new spaces of discourse, none have achieved the mass-mediated prominence of the hashtag. The frictions born from trying to wrest the hashtag away from the hollow articulations of PR and advertising campaigns and maintain its potential for activist content are ongoing. But one upshot of this prominence is, perhaps ironically, that at least in this one way, neoliberalism and activism might be speaking the same language, though obviously with different intents. But might this mean a way out of the familiar cycle of just preaching to the choir in activist circles (Downing & Rodriguez, 2005)? Do political hashtags, despite their inconsistent, somewhat unfocussed, sometimes embattled aspects, have the world's ear? And do race-activist hashtags give us a new tool, not simply for a presumed ineffectual "armchair activism" but for actual discursive intervention?

One final question worth asking is: What is the space between **#RaceFail** and **#Ferguson**? One important thing to note about this gap is that the time between them isn't neutral but rather one of development. The networked persistence of discussions of racism in life and culture (of which the stickiness of **#RaceFail** is a useful metonym) has allowed critical race theory to expand from the bounds of too-insular activist and academic publics into a known apparatus and structure of the public sphere as a whole.[6] The space between these two tags is one of time and growth and development—similar, also, to the space between 9/11 and the Boston bombings or the recent shootings on Parliament Hill in Canada. When the tragedy of 9/11 happened, Western society snapped: racism, always already present and having major structural effects, flooded the public sphere and overwhelmed the critical voices and societal guilt-structures that kept it unevenly and incompletely in check. But in the subsequent events, we had the apparatus of critical race discourse—forged on the fly from the frictions and fractures of that time, combined with earlier writing and activism—to bulwark and embolden our critical voices; intersectional analyses to ask also what roles foreign policy, poverty, and mental health issues might have had in triggering these horrible events; and the sobering memories of the dismal reality of the post-9/11 period to help us resist falling prey to the fearmongering of opportunistic and hawkish politicians. **#Ferguson** and other new and potent race-activist hashtags (such as **#BlackLivesMatter, #FreddieGray**) are the legacy of earlier critical discussions and the earlier technosocial events that mobilized them, **#RaceFail** being but one, and collectively they form an activist critical objection—a quarrel—leveled at unjust abuses of power.

NOTES

1. This is a condensed version of a paper from the "Entanglements: Activism and Technology" special issue of *The Fibreculture Journal*, edited by Pip Shea, Tanya Notley, and Jean Burgess. The full version may be found at http://twentyfive.fibreculturejournal.org. Reprinted by permission.

2. The name and shape of the sign was likewise determined. It is known as the "pound" or "number" sign due to its Roman provenance—derived from the term *libra pondo* (pound in weight), abbreviated "lb"—and evolved through usage and convention to "℔" and finally "#" (Roman, 2014). In the U.K. it is known as the "hash sign" (a corruption of "hatch" from its cross-hatched shape) because their (monetary) pounds have their own distinct symbol. It is also sometimes known as the "octotherp," "octothorp," or "octothorpe," and by other names in further contexts, such as the "sharp" in musical notation—although typographically, this is a similar, though not identical, symbol: "♯"(Roman, 2014).

3. Intimate privilege is the emergent social and cultural privilege that accrues intersectionally to intimacies.

4. September 2013 was when I started to track this tag in real time. For a larger project, I am currently seeking funding to get access to this tag through the now proprietary Twitter API from its emergence in 2009.
5. In a widely criticized move, the grand jury chose not to indict Darren Wilson.
6. See, for example, the mainstream impacts of Black Twitter and its various protest hashtags as discussed by Clark (Chapter 15, this volume), and by Cantey and Robinson (Chapter 16, this volume).

REFERENCES

@haymakerhattie. (2014, April 24). #RaceFail usually means something very different in my circles #AmazingRace [Tweet]. Retrieved from https://twitter.com/HaymakerHattie/status/460579134868316161

@SCRunnerGrl. (2015, February 22). Aaaaaand it's raining #racefail [Tweet]. Retrieved from https://twitter.com/SCRunnerGrl/status/569467101540421633

Ahmed, S. (2009). *Happiness, race, and empire*. Talk presented at International Day for the Elimination of Racial Discrimination, March 19, York University, Toronto.

Al Jazeera. (2014, September 5). US opens Ferguson civil rights inquiry. *Al Jazeera*. Retrieved from http://www.aljazeera.com/news/americas/2014/09/us-opens-ferguson-civil-rights-inquiry-20149422453226832.html

Angry Asian Man. (2014, January 14). What's up with the yellowface on *How I Met Your Mother*? *Angry Asian Man* [Blog]. Retrieved from http://blog.angryasianman.com/2014/01/whats-up-with-yellowface-on-how-i-met.html

Aronofsky, D. (Director). (2014). *Noah* [Film]. Hollywood, CA: Paramount.

Avalon's Willow. (2009, January 13). Open letter: To Elizabeth Bear. *Seeking Avalon* [blog]. Retrieved from: http://seeking-avalon.blogspot.ca/2009/01/open-letter-to-elizabeth-bear.html

Bastos, M. T., Raimundo, R. L. G., & Travitzki, R. (2013). Gatekeeping Twitter: Message diffusion in political hashtags. *Media, Culture and Society, 35*(2), 260–270.

Baym, N. (2010). *Personal connections in the digital age*. Malden, MA: Polity.

Bays, C. (Producer), & Thomas, C. (Producer). (2005). *How I met your mother* [Television series]. Hollywood, CA: CBS.

Bear, E. (2009, January 12). Whatever you're doing, you're probably wrong. *throw another bear in the canoe* [Blog]. Retrieved from http://matociquala.livejournal.com/1544111.html

Beaumont, V. (2014, November 16). Anonymous hijacks KKK Twitter account after Klan declares cyber war (screenshots). *IfYouOnlyNews.com*. Retrieved from http://www.ifyouonlynews.com/racism/anonymous-hijacks-kkk-twitter-account-after-klan-declares-cyber-war-screenshots/

Benioff, D. (Producer), & Weiss, D. B. (Producer). (2011). *Game of thrones* [Television series]. Hollywood, CA: HBO.

Benovitz, M. G. (2010). "Because there aren't enough spoons": Creating contextually-organized argument through reconstruction. *National Communication Association Annual Conference, Conference Proceedings* (pp. 124–130). Washington, DC: National Communication Association.

Berlant, L. (1998). Intimacy: A special issue. *Critical Inquiry, 24*(2), 281–288.

Bruns, A. (2012). How long is a tweet? Mapping dynamic conversation networks on Twitter using Gawk and Gephi. *Information, Communication and Society, 15*(9), 1323–1351.

Bruns, A., & Burgess, J. (2012). Researching news discussion on Twitter: New methodologies. *Journalism Studies, 13*(5–6), 801–814.

Chee, F., & Bergstrom, K. (2013). *On playing "like a girl": A comparative analysis of quasi-affirmative (re)action.* Paper presented at Canadian Communication Association, June 5–7, Victoria, BC.

Covert, B. (2014, October 13). Largest Ferguson demonstration yet ends with sit-in. *Think Progress.* Retrieved from http://thinkprogress.org/justice/2014/10/13/3579229/ferguson-sit-in/

Digital Spy. (2014, January 15). *How I Met Your Mother* creators apologize for Yellowface incident. *Cosmopolitan.* Retrieved from http://www.cosmopolitan.com/entertainment/tv/news/a18635/how-I-met-your-mother-apologizes-for-yellowface/

Doganieri, E. (Producer), & van Munster, B. (Producer). (2001). *The amazing race* [Television series]. Hollywood, CA: CBS.

Douai, A. (2013). "Seeds of change" in Tahrir Square and beyond: People power or technological convergence? *American Communication Journal, 15*(1), 24–33.

Downing, J., & Rodriguez, C. (2005). *Citizens' media: Methods, problems and prospects.* Keynote address presented at Media Democracy Day, October 18, Concordia University, Montréal.

Fairbanks, C. (2014, November 12). Missouri governor says "violence will not be tolerated" in Ferguson, a viral hashtag is born. *The Free Thought Project.* Retrieved from http://thefreethoughtproject.com/missouri-governor/#ur3MJ5KdkqbaMTbL.99

Foucault, M. (1972). Discourse on language. In A. M. Sheridan Smith (Trans.), *The archaeology of knowledge and the discourse on language* (pp. 215–237). New York: Pantheon.

Foucault, M. (1990). *The history of sexuality, volume 1: An introduction* (Trans. R. Hurley). New York: Vintage. (Original work published 1976)

Fraser, N. (1992). Rethinking the public sphere: A contribution to the critique of actually existing democracy. In C. Calhoun (Ed.), *Habermas and the public sphere* (pp. 109–142). Cambridge, MA: MIT Press.

Habermas, J. (1984). *The theory of communicative action, vol. 1, Reason and the rationalization of society* (Trans. T. McCarthy). Boston: Beacon Press. (Original work published 1981)

Habermas, J. (1989). *The structural transformation of the public sphere: An inquiry into a category of bourgeois society* (Trans. T. Burger & F. Lawrence). Cambridge, MA: MIT Press. (Original work published 1962)

Hathaway, J. (2014, October 10). What is Gamergate, and why? An explainer for non-geeks. *Gawker.* Retrieved from http://gawker.com/what-is-gamergate-and-why-an-explainer-for-non-geeks-1642909080

Hopkinson, N. (2010). A reluctant ambassador from the planet of Midnight. *Journal of the Fantastic in the Arts, 21*(3), 339–350.

Huang, S.-L. (Director), & Rosenblum, J. (Director). (2013). *The naked brand* [Documentary]. New York: Questus.

Jemisin, N. K. (2010, January 18). Why I think RaceFail was the bestest thing evar for SFF. *Epiphany 2.0* [Blog]. Retrieved from http://nkjemisin.com/2010/01/why-i-think-racefail-was-the-bestest-thing-evar-for-sff/

Jemisin, N. K. (2013, June 8). Continuum GoH speech. *Epiphany 2.0* [Blog]. Retrieved from http://nkjemisin.com/2013/06/continuum-goh-speech/

Kroh, K. (2014, 10 August). Missouri town erupts in protest after police shoot unarmed black teenager. *Think Progress.* Retrieved from http://thinkprogress.org/justice/2014/08/10/3469602/ferguson-police-michael-brown/

Kwinter, S. (2001). The complex and the singular. In *Architectures of time: Toward a theory of the event in modernist culture* (pp. 3–31). Cambridge, MA: MIT Press.

Massumi, B. (2011). *Semblance and event: Activist philosophy and the occurrent arts.* Cambridge, MA: MIT Press.

Messina, C. (2007, August 25). Groups for Twitter; or a proposal for Twitter tag channels. *FactoryCity* [Blog]. Retrieved from http://factoryjoe.com/blog/2007/08/25/groups-for-twitter-or-a-proposal-for-twitter-tag-channels/

Meyer, R. (2014, June 28). Everything we know about Facebook's secret mood manipulation experiment. *The Atlantic*. Retrieved from http://www.theatlantic.com/technology/archive/2014/06/everything-we-know-about-facebooks-secret-mood-manipulation-experiment/373648/

Mosco, V. (2004). *The digital sublime: Myth, power, and cyberspace*. Cambridge, MA: MIT Press.

Papacharissi, Z., & Oliveira, M. de F. (2012). Affective news and networked publics: The rhythms of news storytelling on #Egypt. *Journal of Communication, 62,* 266–282.

Peters, J. D. (1999). *Speaking into the air: A history of the idea of communication.* Chicago: University of Chicago Press.

Petersen, K. (2014, August 20), Unrest in Ferguson draws attention from the US's critics. *BBC News.* Retrieved from http://www.bbc.com/news/blogs-echochambers-28845076

Racialicious. (2014, January 15). [Untitled post]. *Tumblr.* Retrieved from http://racialicious.tumblr.com/post/73432116354/as-if-you-didnt-have-enough-of-a-reason-to-hate

Rambukkana, N. (2015). *Fraught intimacies: Non/monogamy in the public sphere.* Vancouver, BC: UBC Press.

Roman. (2014, December 16). Episode 145: Octothorpe. *99% invisible* [Online radio program]. Retrieved from http://99percentinvisible.org/episode/octothorpe/

Ross, G. (Director). (2012). *The hunger games* [Film]. Hollywood, CA: Lionsgate.

Shyamalan, M. N. (Director). (2010). *The last airbender* [Film]. Hollywood, CA: Paramount.

Smietana, B. (2014, August 4). Does "Noah" have a race problem? Biblical film draws criticism for lack of diversity. *Huffington Post.* http://www.huffingtonpost.com/2014/04/08/noah-race_n_5107490.html

Somerville, A. (n.d.). A themed summary of RaceFail '09 in large friendly letters for those who think race discussions are hard. *Fiction by Ann Somerville: Love, romance, and the occasional sound thrashing* [Blog]. Retrieved from http://annsomerville.net/a-themed-summary-of-racefail-09-in-large-friendly-letters-for-those-who-think-race-discussions-are-hard/

Starr, T. J. (2014, August 17). Woman behind powerful Mike Brown protest photo defies "respectability politics." *AlterNet.* Retrieved from http://www.alternet.org/woman-behind-powerful-mike-brown-protest-photo-defies-respectability-politics

Townes, C. (2014, November 12). "Ready for war": 1,000 police officers mobilized in advance of grand jury ruling in Ferguson. *ThinkProgress.* Retrieved from http://thinkprogress.org/justice/2014/11/12/3591240/ferguson-grand-jury-ruling-prep/

Tsing, A. L. (2005). *Friction: An ethnography of global connection.* Princeton, NJ: Princeton University Press.

Tufekci, Z. (2014, August 14). What happens to #Ferguson affects Ferguson: Net neutrality, algorithmic filtering and Ferguson. *Medium.com.* Retrieved from https://medium.com/message/ferguson-is-also-a-net-neutrality-issue-6d2f3db51eb0

Warner, M. (2002). *Publics and counterpublics.* New York: Zone Books.

Whitehead, A. N. (1967). *Process and reality* (corrected ed.). New York: Free Press.

Zappavigna, M. (2011). Ambient affiliation: A linguistic perspective on Twitter. *New Media & Society, 13*(5), 788–806.

Zerehi, S. S. (2014, August 19). Michael Brown's shooting in Ferguson lost on social media. *CBC News.* Retrieved from http://www.cbc.ca/news/technology/michael-brown-s-shooting-in-ferguson-lost-on-social-media-1.2740014

#auspol: The Hashtag as Community, Event, and Material Object for Engaging with Australian Politics

THERESA SAUTER AND AXEL BRUNS

INTRODUCTION

Critical, loud, highly discursive and polarised, the **#auspol** hashtag represents a space, an event and a network for politically involved individuals to engage in and with Australian politics and perform political participation and communication. As a long-standing institution in the Australian Twittersphere (see, e.g., Bruns & Burgess, 2011; Bruns & Stieglitz, 2012, 2013), the **#auspol** hashtag provides a potent case study through which to explore the material, relational and discursive dimensions of a hashtag public. This chapter engages with the use of this particular hashtag, both empirically and theoretically. In particular, we work through a number of models that can be used to characterise the **#auspol** hashtag: it is, at different times and even at once, a discursive community of users; a mechanism for tracking and engaging in specific political events; and an object of discussion and controversy in its own right.

We use our long-term study of the **#auspol** community as a case study for considering how hashtags can mediate public engagement with politics. In this way, we conceptualise the hashtag as an everyday material object that contributes to the unfolding of social and political reality. Tools such as hashtags are objects that are embedded in and materialise out of shared interests, issues and events. At the same time as they emerge out of such shared experiences, hashtags are also involved in shaping them. In this way, the hashtag is an object, event and relational encounter that can transform political participation.

This perspective draws on recent work in the area of material participation (Marres, 2012; Michael, 2012) and science and technology studies (STS) (Fuller, 2011; Woolgar & Lezaun, 2013), as well as on more established conceptualisations of the role of nonhumans, objects and events as relational concepts that are intimately entwined in processes of shaping publics (Deleuze, 2003; Foucault, 1972; Latour, 1999, 2005; Whitehead, 1929, 1933). This is not to argue that Twitter, its hashtags or other material objects and events heighten or dampen, promote or prevent public engagement. Rather, they play an active role in the unfolding of political realities and thus need to be acknowledged and understood as more than mere communicative markers.

Bruns and Moe (2014) provide a useful framework for thinking about the various structural layers of communication on Twitter. They show that hashtags coordinate exchanges around specific topics, issues or events at the 'macro level' of communication. Hashtags complement quasi-private @reply conversations (at the micro level) and flows of information across follower-followee networks (at the meso level) by enabling the gathering and interactions of much broader, more visible and dynamic publics comprised of users who need not follow or even be aware of each other prior to their participation in the hashtag; such publics often come together and disperse ad hoc (see also Bruns & Burgess, 2011; Chapter 1, this volume). Notably, not all hashtags operate this way: Some hashtags (like #win, #fail or #facepalm) are mainly used as paratextual markers akin to emoticons or punctuation marks. However, an important subset of all hashtags, including #auspol, are used to mark out a specific discursive territory and facilitate the coming together of participants with shared thematic interests. Even such thematic hashtags, however, vary in their uses, depending especially on the inherent dynamics of the theme they are designed to address.

Studying the dynamics of a long-standing thematic hashtag community such as #auspol provides important insights into the hybridity of the hashtag as both topical marker and discursive technology. In this chapter, we explore the role of the #auspol hashtag as a discursive marker of the relationships of participants to other community members, as a mechanism for tracking and discussing unfolding political events and as an object that marks a particular form of communicative exchange within the wider context of public debate. We thus 'flatten out' (Latour, 2005) the concept of the thematic hashtag, accounting for it as more than a simple communicative tool while avoiding any claims that it is itself determinant of political reality. Hashtags are thus neither fully material nor fully symbolic but rather exercise an important agency in the construction of power relations, events and knowledge. They are tied up with other objects, subjects, contexts, events, relations, discourses and truths that extend well beyond the specific context of Twitter as a platform, or even the Internet more generally. To apply a basic Latourian line of argument, hashtags are actants in a network that 'do things'. They are one point that allows us to trace more extensive connections within an (infinite) network of actors.

THE #AUSPOL HASHTAG

While its precise origins are by now difficult to retrace (Twitter's Application Programming Interface does not provide the functionality to identify the first tweets to long-standing hashtags), **#auspol** clearly is one of the oldest and best-established hashtags in the Australian Twittersphere. Its comparative volume and popularity reflect both the overall demographics of the Australian Twitter user base—which remain skewed to a subset of the population that is especially politically active—and the fact that Australian politics has experienced an unusually turbulent period since the 2007 federal election, with three changes to the prime ministership between 2010 and 2013 alone. A preliminary map of follower/followee relationships in the overall Australian Twittersphere which we developed in 2012 (Bruns, Burgess, & Highfield, 2014; Bruns, Burgess, Kirchhoff, & Nicolai, 2012) shows a significant portion of the network to be structured by a shared interest in news and politics (Figure 3.1a), with participants in the **#auspol** hashtag predominantly recruited from the same areas of the overall network (Figure 3.1b).

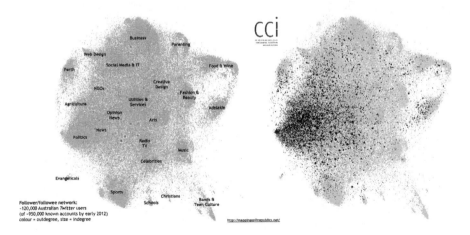

Figure 3.1. a) Overall map of follower/followee relationships in the Australian Twittersphere, as of 2012; b) Overall map (light grey), with most active participants in **#auspol** highlighted (dark grey).

Further quantitative analysis of **#auspol** activities, following the methodologies outlined by Bruns and Stieglitz (2012, 2013), documents the very substantial overall volume of tweets, and reveals a number of key distinguishing features. Such activities are necessarily influenced by day-to-day political events. In our analysis, we focus first on a (comparatively) stable period in Australian politics during 2011, which saw neither a challenge to or change of leadership nor a federal election campaign. For this period, we draw on data gathered from the start of February 2011 to the start of December 2011, thus avoiding the comparatively slower summer holiday months of December and January.

During this time, we captured over 850,000 **#auspol** tweets, posted by some 26,000 unique accounts—an average of some 85,000 tweets per month, or 2,800 tweets per day. More important even than this very significant volume of activity is the distribution of participation across the **#auspol** contributor base: of the overall 850,000 tweets, over 550,000 (more than 64%) were contributed by the most active 1% of the overall user base—that is, by some 265 unique accounts in total (Figure 3.2). Combined with a second group of still highly active users, the top 10% of **#auspol** contributors account for more than 91% of the total volume of **#auspol**. On average, the 265 members of the leading group each posted more than 2,000 **#auspol** tweets during the 10-month period examined here, in other words—but in reality the distribution is even more concentrated around a very small group of lead users: seven leading participants each posted more than 10,000 **#auspol** tweets during the 10 months examined here.

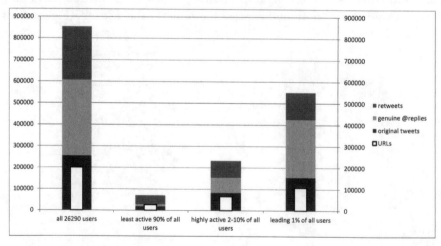

Figure 3.2. Participation patterns in **#auspol**, early February to early December 2011.

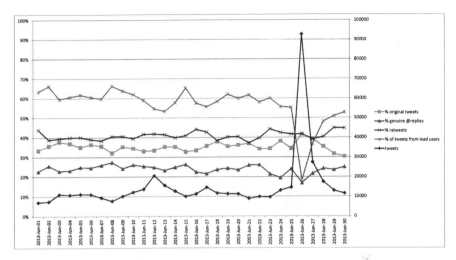

Figure 3.3. Participation patterns in **#auspol**, June 2013.

#AUSPOL AS COMMUNITY: HASHTAGS
AS RELATIONAL MARKERS

It is also notable from Figure 3.2 that a relative majority of tweets to **#auspol** are genuine @replies rather than retweets or original tweets (i.e., tweets which do not specifically address any other user). This is especially pronounced for the lead group, for whom 49% of all tweets posted are genuine @replies. Although not all such @replies will be directed to other members of the lead group or to other **#auspol** participants (they may also @mention politicians' or journalists' Twitter accounts, for example), this nonetheless points to a very strong conversational element in **#auspol** engagement.

This existence of a highly active, interactive and concentrated group of lead users at the heart of **#auspol** points strongly to a conceptualisation of **#auspol** as a tightly knit community of participants who—in spite of their possible political differences—share a common commitment to discussing Australian politics. From this perspective, the persistent use of a hashtag like **#auspol** in one's tweets signifies participation—as a Twitter user in a particular community of debaters, and more generally as a contributor to public deliberation around Australian politics. Viewed this way, then, the hashtag is a relational marker. Adding a hashtag to a tweet signals the desire for the message to be seen by a particular public cohort and thereby calls for interaction. Hence, the hashtag may be described as a 'technology of engagement' (Marres, 2012, p. x): a tangible, material tool for people to relate to themselves and to others.

Applying a Foucauldian understanding of 'technology' to Marres's terminology, the hashtag can be conceptualised as a 'technique of the self' (Foucault, 1988): a practice of relating to self and others in order to develop ethical guidelines for governing one's life. Contributing to a hashtag discussion on Australian politics—or any other discursive topic—is a mundane way of engaging with the political context and one's own reactions to and understandings of it, as well as those of others who experience the same occurrences, in order to make sense of and navigate through it. By using the **#auspol** hashtag, Twitter users come to understand and participate in the everyday realities they are enrolled in and thus establish their roles within them in mundane and perhaps unconscious ways. Furthermore, the use of a hashtag also flags the desire to participate and perform in a community, shaping further ways of relating to and forming oneself.

Yet the very concentrated structure of the **#auspol** lead user group, combined with the network structure shown in Figure 3.1, also complicates the conversational and relational nature of its use. Figure 3.1b shows the most active contributors to **#auspol** to be located in a tight network cluster on the sidelines of the overall 'news and politics' network within the Australian Twittersphere (at the centre left of the graph)—to the extent that these lead users are conversing amongst one another, then, there is very little actual need for the functionality which a hashtag like **#auspol** provides: they are already using @replies to directly address each other and could otherwise rely on their existing mutual follower/followee relationships, which ensure that all tweets posted by its members are visible to this lead group. We suggest that using **#auspol** in this particular discursive space has a strong performative aspect: the lead group's **#auspol** conversations are performed, somewhat in the style of a podium debate, to an imagined audience of other **#auspol** users who follow but only occasionally actively engage in the discussion (cf. Pearson, 2009, on performance on SNSs).

Employing a thematic hashtag like **#auspol** is therefore also a way of navigating power relations. Although newly created hashtags and their pools of contributors (and indeed, any new space for online participation) may start out comparatively unformed and unstructured, it is virtually inevitable that continued engagement by contributors who differ in their levels of activity, expertise and commitment leads to a stratification of the user base and to the formation of what may be described as community structures—indeed, to the transformation of a user base into a community in the full sense of that term (cf. Katz et al., 2004; Rheingold, 1993). This structured community—which clearly exists in the case of **#auspol**—consists in the first place of a network of power relationships between individual contributors, which are established and maintained through continued participation. New participants must read and understand this network of relationships in order to take part in the community. Viewed in this way, the thematic hashtag must be seen as a relational marker, enrolled in and constitutive of activities that are characterised by engagements with self and others and by the navigation of power relations.

#AUSPOL DURING #SPILL: HASHTAGS AND/AS EVENTS

Thematic hashtags rarely exist in isolation from a wider discursive context: they emerge out of, are part of and can shape events in the wider online and offline world. Most fundamentally, without its embedding in the wider Australian sociopolitical context, the #auspol hashtag could not exist and be sustained. The dynamics of hashtag conversations and communities are influenced by day-to-day issues and events—in our example, the unfolding story of Australian politics at the federal level.

Although the #auspol hashtag is usually dominated—by way of sheer volume—by a small number of participants whose activities serve to drown out most other voices, at times of heightened political attention, these lead users may themselves be overwhelmed by a temporary influx of new participants who are drawn to #auspol by the drama of unfolding events without bothering to understand and negotiate existing power relations within this discursive space. To illustrate this momentary reversal of community structures within the hashtag space, we draw on an especially incisive moment in recent Australian political history: the 2013 Australian Labor Party (ALP) leadership change.

In 2007, ALP leader Kevin Rudd had won a decisive election victory over conservative Prime Minister John Howard; with opinion polls softening, however, Rudd's party took the extraordinary step of replacing him with his deputy Julia Gillard in mid-2010, only months before the next federal election in November 2010—an event known in Australian political parlance as a 'leadership spill'. Gillard won the subsequent election, narrowly, and with the support of two independent members of parliament formed government but continued to be affected by poor opinion polls and sustained opposition attacks on her character. In turn, her predecessor Kevin Rudd challenged Gillard to regain the ALP leadership as well as the prime ministership—once, unsuccessfully, in February 2012, and again, successfully, on 26 June 2013. (Rudd lost the subsequent federal election to conservative opposition leader Tony Abbott in September 2013.)

Each spill caused considerable activity in the Australian Twittersphere. While a substantial component of such activity takes place in a dedicated #spill hashtag first used during the 2010 leadership challenge, #auspol—as the nominally standard hashtag for everyday political discussion in Australia—is also significantly affected by the event. For the purposes of the following discussion, we examine **#auspol** activities during June 2013 (Figure 3.3).

It is immediately evident that there is a substantial shift in the dynamics of **#auspol** on 26 June, continuing to a lesser extent on subsequent days. While the average volume of posts (shown in Figure 3.3 as a dashed line) hovers just above 11,000 tweets per day during the first 25 days of the month, it shoots to more than 92,000 tweets on the day of the spill and gradually returns to the baseline level by 29 June. Conversely, the average volume of tweets contributed by the most active 1% of users (calculated over the entire month) sits at 60% during these first weeks of the month and thus remains at a level comparable to the 64% we saw during 2011—yet their contribution to the total **#auspol** volume drops to only 18% on 26 June and recovers only gradually to just over 50% over the remainder of the month. Notably, 26 June also records the monthly minimum for the percentage of tweets of the total **#auspol** volume which are genuine @ replies.

In combination, these observations speak of a considerable if temporary power shift within the **#auspol** community: far from its usual domination by a handful of highly active leading users, participation during the heady days of the spill and its aftermath is substantially more broadly based. This shift, then, marks **#auspol**'s temporary transition from community to event: where usually its main function is to sustain the maintenance of power relationships between its more or less active participants, it now predominantly serves to support the continuing tracking and evaluation of an unfolding political crisis as driven by a much larger, much more fluid ad hoc gathering of participants. In other words, what can occur under such circumstances is a shift from the static to the dynamic and from the spatial to the temporal. As such, it holds the potential to disrupt and reshape the existing status quo of power relations within **#auspol**: the community power structures which re-ossify once the immediate crisis is over may well differ from those which have existed before, if Twitter activities around the spill event have provided new participants with a platform to prove themselves or if existing lead users have failed to keep pace with unfolding events. Indeed, disruptive events such as the leadership spill and its coverage on Twitter could even challenge the implicit primacy of **#auspol** for Australian political discussion on Twitter if they generate new, widely known hashtags that manage to survive beyond the immediate event itself.

But thematic hashtags such as **#auspol** are never entirely removed from events in the wider world, even when there is no major political crisis unfolding. Deleuze (1990, p. 8), with reference to Whitehead, proposes that the event 'is always that which has just happened and that which is about to happen, but never that which is happening' (cf. Latour, 1999). Even when **#auspol** functions mainly as a mechanism to sustain a community of participants, the currency of that community is its discussion of continuing political events in Australia, and the units of that currency—the individual contributions made by community members—are microscopic events in their own right. Hence, to understand the hashtag both as accompanying events and as itself comprising events requires us to acknowledge the interconnectedness of online and offline occurrences. What changes in the shifts between hashtag-as-community and hashtag-as-event is the valuation of that currency and its pegging to outside occurrences—as a result, the hashtag is revealed as a highly malleable discursive object which is embedded in a wider network of interrelationships.

VOX TWITTERATORUM: HASHTAGS AS OBJECTS

An important strand of research has emphasised the role of object ecologies and material culture in the shaping of daily practices, interactions and networks (cf. Appadurai, 1986; Dant, 2005, 2008; Knorr Cetina, 2001; Latour, 2005; Rambukkana, Introduction, this volume), and considered what constitutes an 'object' in a digital context (cf. Leonardi, 2010; Leonardi, Nardi, & Kallinikos, 2012; Marres, 2012). Leonardi (2010) discusses how digital artifacts (in spite of their intangibility) provide certain affordances and constraints that limit and direct how we use them, much as material objects like wood, glass, metal, etc. come with certain possibilities and limitations to how they can be used. Hence, while a topical hashtag such as **#auspol** is used as a communicative marker that signals someone's desire to contribute to a conversation around a particular topic, it also has a material tangibility in itself as a distinctive contributor to an assemblage of human and nonhuman agents that constitute the conversation. The hashtag can thus be considered a 'participatory object' (Marres, 2012, p. 9), both in terms of its ability to engage Twitter users in discussion (as can be seen from the long-standing and ongoing interaction via the **#auspol** hashtag) and as a participant in this political discussion itself.

Our analysis of the use of the **#auspol** hashtag during the June 2013 Australian Labor Party leadership spill (Sauter & Bruns, 2013a) showed that the hashtag was employed in messages that expressed support of or discontent with political parties and politicians. Importantly, the conversation would not exist without an external context—in our case, the leadership spill—yet at the same time, the materiality of the hashtag itself makes it possible for the conversation to be framed and made visible to a particular ad hoc public. Just as we think about the users, the computers, tablets, phones, or other electronic devices they use to compose their tweets and the politicians and sociopolitical circumstances they tweet about as having a material tangibility, we have to recognise the **#auspol** hashtag as also contributing an element to the way in which the discussion unfolds and is perceived and shaped. Hence, even though it is an electronically constituted actant without any physical materiality in a literal sense, its presence in the discussion of the topic affects the discussion, making it a fully realised actor well beyond the confines of Twitter itself.

This is perhaps most obvious in the context of unfolding events: during each of the various attempted and actual Labor leadership spills since 2010, for example, the increased volume of **#auspol** activity and the coming into existence of **#spill** and related event-specific hashtags was seen by everyday users, political journalists and commentators and even politicians themselves as an indicator that a new leadership challenge 'is on'. This public attention to **#spill** and related hashtags may be a playful, mischievous, or calculating appropriation of Twitter activity patterns which widely overstates the actual influence of Twitter on Australian political processes. However, the very fact that such (mis)appropriation of Twitter hashtags is even possible demonstrates the materiality of the thematic hashtag as an object in contemporary Australian politics. The materiality of the **#spill** hashtags on Twitter and elsewhere gives commentators licence to explore the possibility of a new spill.

But outside of such specific, rarefied events, too, the very material presence of **#auspol** as a thematic hashtag creates the mechanism through which an aggregate 'vox Twitteratorum' (in analogy to the similarly aggregate vox populi) can be incorporated into public debate. The creation of such a disembodied 'voice of the Twitterati' and its operationalisation in the depiction of political discourse becomes evident, for example, from studies of how Twitter is increasingly cited as a source of views and comments by political journalists as they cover current political issues and debates (cf. Sauter & Bruns, 2013b; Wallsten, forthcoming). Because of its participant structure, **#auspol**, and Twitter, is highly unlikely to be representative of Australian public opinion—but some political journalism now positions it as such.

As Leonardi (2010) puts it, then, 'artifacts without matter, matter'. Hashtags such as **#auspol** and **#spill** made everyday political debate on Twitter, the communities conducting such debates and the ad hoc discussion of the various leadership spills and other events visible to a wider public as well as to us as researchers, imbuing such debates with further meaning and context. (In return, our various published analyses of **#spill**, **#auspol** and other relevant hashtags have also contributed to increasing their visibility and prominence—researchers and their publications are themselves also actors and actants in the network, of course.) Therefore, a thematic hashtag is a 'participatory object' in a concrete sociopolitical context. As an object with a palpable materiality, it is able to convene a conversation and facilitate its visibility, and in doing so, it also shapes what people talk about and how they do so.

CONCLUSION: HASHTAGS IN CONTEXT

Finally, what must not be forgotten in this context is the interplay between thematic hashtags as discursive technologies as such and the more fundamental technological systems upon which they are founded: Twitter's underlying software base and the algorithms inscribed into it. Gillespie (2014) alerts us to the ways in which algorithms increasingly shape what we know or think is worth knowing and how we are known. With reference to Langlois, he asserts that algorithms are 'a key logic governing the flows of information on which we depend, with the power to "enable and assign meaningfulness, managing how information is perceived by users"' (Langlois, 2012, cited in Gillespie, 2014, p. 167). Similarly, hashtags are implicated in algorithmic processes of categorising and visualising information and determining access to such knowledge.

By adding the **#auspol** hashtag to a tweet, a user makes a decision to trigger an algorithm in the Twitter software base, which associates the tweet—and, by extension, the user—with a particular topic and group of participants. In this way users add to a publicly visible body of data: they contribute to the negotiation of truth via public debate and thus participate in the construction of knowledge. Depending on context, this may be done consciously—as expressed most obviously in efforts to get a certain hashtag to 'trend' and thus to afford greater visibility to the event or issue associated with it—or inadvertently, when a hashtag is used merely as a routine way of referring to a specific issue or theme, without the user necessarily seeking to engage explicitly with the community that may exist around the hashtag.

The thematic hashtag's agency should not be overestimated, however. It is one participant in the way communicative exchanges unfold; many other tangible and intangible, more or less powerful participants are similarly involved in such processes. We have to remain aware that what we are observing when we trace the use of a hashtag is a particular snapshot of a very specific group of participants. Furthermore, we cannot trace all of the elements that impacted on the activities we observe through the hashtag, such as the conversations that the contributors to a hashtag public such as **#auspol** have beyond the use of the hashtag itself, whether on Twitter or elsewhere, and how this affects the ways in which they present and relate to themselves and others and engage in the construction of knowledge. The thematic hashtag is thus only one discursive technology at the macro-level of Twitter communication; many more such technologies—from other, nonthematic hashtags to the various other means for communication at the meso and micro levels—exist alongside and in competition with it on Twitter alone. A number of other technologies which is greater still by an order of magnitude exists beyond the narrow confines of Twitter itself. Studying the particular affordances and implications of hashtagged discourse on Twitter thus also becomes an exercise in mapping one very specific discursive space. This exercise should be repeated for a great many other discursive spaces—on Twitter and beyond—so that these spaces may be better positioned and understood in relation to each other.

REFERENCES

Appadurai, A. (1986). Introduction: Commodities and the politics of value. In A. Appadurai (Ed.), *The social life of things: Commodities in cultural perspective* (pp. 3–63). Melbourne: Cambridge University Press.

Bruns, A., & Burgess, J. (2011). *The use of Twitter hashtags in the formation of ad hoc publics.* Paper presented at the 6th European Consortium for Political Research General Conference, August 25–27, University of Iceland, Reykjavík.

Bruns, A., Burgess, J., & Highfield, T. (2014). A 'big data' approach to mapping the Australian Twittersphere. In K. Bode & P. Arthur (Eds.), *Advancing digital humanities* (pp. 113–120). Basingstoke, UK: Palgrave Macmillan.

Bruns, A., Burgess, J., Kirchhoff, L., & Nicolai, T. (2012). *Mapping the Australian Twittersphere.* Paper presented at Digital Humanities Australasia, March 30, Canberra.

Bruns, A., & Moe, H. (2014). Structural layers of communication on Twitter. In K. Weller, A. Bruns, J. Burgess, M. Mahrt, & C. Puschmann (Eds.), *Twitter and society* (pp. 29–41). New York: Peter Lang.

Bruns, A., & Stieglitz, S. (2012). Quantitative approaches to comparing communication patterns on Twitter. *Journal of Technology in Human Services, 30*(3–4), 160–185.

Bruns, A., & Stieglitz, S. (2013). Towards more systematic Twitter analysis: Metrics for tweeting activities. *International Journal of Social Research Methodology, 16*(2), 91–108.

Dant, T. (2005). *Materiality and society.* Buckingham, UK: Open University Press.

Dant, T. (2008). The pragmatics of material interaction. *Journal of Consumer Culture, 8*(1), 11–33.

Deleuze, G. (1990). *The logic of sense* (M. Lester & C. Stivale, Trans., C. Boundas, Ed.). London: Athlone Press.

Deleuze, G. (2003). *The fold: Leibniz and the baroque.* London & New York: Continuum.

Foucault, M. (1972). *The archaeology of knowledge and the discourse on language.* London: Tavistock Publications.

Foucault, M. (1988). Technologies of the self. In L. H. Martin, H. Gutman, & P. H. Hutton (Eds.), *Technologies of the self—A seminar with Michel Foucault* (pp. 16–49). London: Tavistock Publications.

Fuller, S. (2011). *Humanity 2.0: What it means to be human past, present and future.* London: Palgrave Macmillan.

Gillespie, T. (2014). The relevance of algorithms. In T. Gillespie, P. Boczkowski, & K. Foot (Eds.), *Media technologies: Essays on communication, materiality, and society* (pp. 167–194). Cambridge, MA: MIT Press.

Katz, J. E., Rice, R. E., Acord, S., Dasgupta, K., & Kalpana, D. (2004). Personal mediated communication and the concept of community in theory and practice. *Communication Yearbook, 28,* 315–372.

Knorr Cetina, K. (2001). Objectual practice. In T. R. Schatzki, K. Knorr Cetina, & E. von Savigny (Eds.), *The practice turn in contemporary theory* (pp. 175–188). London: Routledge.

Latour, B. (1999). *Pandora's hope: Essays on reality of science studies.* Cambridge, MA & London: Harvard University Press.

Latour, B. (2005). *Reassembling the social.* Oxford, UK: Oxford University Press.

Leonardi, P. M. (2010). Digital materiality? How artifacts without matter, matter. *First Monday, 15*(6). Retrieved from http://firstmonday.org/ojs/index.php/fm/article/view/3036/2567

Leonardi, P. M., Nardi, B. A., & Kallinikos, J. (Eds.). (2012). *Materiality and organizing.* Oxford, UK: Oxford University Press.

Marres, N. (2012). *Material participation: Technology, the environment and everyday publics.* Houndmills, UK & New York: Palgrave Macmillan.

Michael, M. (2012). De-signing the object of sociology: Toward an 'idiotic' methodology. *The Sociological Review, 60,* 166–183.

Pearson, E. (2009). All the World Wide Web's a stage: The performance of identity in online social networks. *First Monday, 14*(3). Retrieved from http://firstmonday.org/article/view/2162/2127

Rheingold, H. (1993). *The virtual community: Homesteading on the electronic frontier.* Reading, MA: Addison-Wesley.

Sauter, T., & Bruns, A. (2013a). *Exploring emotions on #auspol: Polarity, and public performance in the Twitter debate on Australian politics.* Paper presented at the 14th Annual Conference of the Association of Internet Researchers, October 23–26, Denver.

Sauter, T., & Bruns, A. (2013b). *Social media in the media: How Australian media perceive social media as political pools.* ARC Centre of Excellence for Creative Industries and Innovation. Retrieved from http://cci.edu.au/socialmediainthemedia.pdf

Wallsten, K. (forthcoming). "New media" in the newsroom: Twitter as a news source during the 2012 campaign. *Newspaper Research Journal.*

Whitehead, A. N. (1929). *Process and reality: An essay in cosmology.* New York: The Free Press.

Whitehead, A. N. (1933). *Adventures of ideas.* Cambridge, UK: Cambridge University Press.

Woolgar, S., & Lezaun, J. (2013). The wrong bin bag: A turn to ontology in science and technology studies? *Social Studies of Science, 43*(3), 321–340.

CHAPTER FOUR

Hashtag as Hybrid Forum:
The Case of #agchatoz

JEAN BURGESS, ANNE GALLOWAY, AND THERESA SAUTER

INTRODUCTION: HASHTAG PUBLICS, HYBRID FORUMS

This chapter imports Michel Callon's model of the 'hybrid forum' (Callon, Lascoumes, & Barthe, 2009, p. 18) into social media research, arguing that certain kinds of hashtag publics can be mapped onto this model. It explores this idea of the hashtag as hybrid forum through the worked example of #agchatoz—a hashtag used as both 'meetup' organiser for Australian farmers and other stakeholders in Australian agriculture, and as a topic marker for general discussion of related issues. Applying the principles and techniques of digital methods (Rogers, 2013), we employ a standard suite of analytics to a longitudinal dataset of #agchatoz tweets. The results are used not only to describe various elements and dynamics of this hashtag but also to experiment with the articulation of such approaches with the theoretical model of the hybrid forum, as well as to explore how controversies animate and transform such forums as part of the emergence and cross-pollination of issue publics.

We proceed on the understanding that publics are multiple and emergent—that is, they are constituted through their material involvement with issues and events rather than pre-existing as a 'public sphere' (Marres, 2012). Digital media platforms are transforming both the nature of such publics and the means through which they engage with issues (Papacharissi, 2010; Bruns & Burgess, Chapter 1, this volume); digital methods present significant new opportunities to observe, describe and understand how issue publics work. In digital media and communication research, it has so far been Twitter that has been most visibly the target of digital methods-based approaches to studying public communication in this way—see for example the chapters collected in *Twitter and Society* (Weller, Bruns, Burgess, Puschmann, & Mahrt, 2014).

Within such approaches, hashtags are often used to focus empirical research on the dynamics of public communication, on a range of traditional topics extending from elections to natural disasters and television audiences (Bruns & Burgess, Chapter 1, this volume; Bruns & Stieglitz, 2012; Deller, 2011), arguably leading to a saturation of what we might call 'hashtag studies'. Most such studies consider the hashtag to be a pragmatic communicative marker that serves to coordinate discussions and establish more or less stable and consistent groups of contributors (see Halavais, 2013, for an overview of the early applications of hashtags, or 'channel tags' as they were dubbed at first). Bruns and Moe (2013) highlight how hashtags generally operate at the macro level of communication on Twitter (with follower-followee networks being at the meso level and @replies at the micro level).

Bruns and Stieglitz (2012) differentiate between three different types of hashtags: *ad hoc* ones, which emerge 'in response to breaking news or other unforeseen events' (p. 165); *recurring* ones, which users employ to contribute repeatedly to a certain topic (such as **#agchatoz**, which we investigate in this chapter); and *praeter hoc* ones, which relevant organisations predetermine and encourage users to adopt when tweeting about a particular event, such as a conference or TV show. Bruns and Moe (2013) further distinguish between topical and nontopical hashtags. They suggest that *topical* hashtags are used to contribute to a discussion on a particular topic. These can be long-standing themes (e.g., **#auspol**), backchannels to TV events (e.g., **#masterchef**), or reactions to particular issues or events (**#royalwedding**). *Nontopical* hashtags such as **#facepalm** or **#fail** are emotive markers and can be applied to any type of tweet. Nontopical hashtags are a deviation from the initially intended use of the hashtag, yet they still serve a communicative purpose.

While hashtags may have originally been intended as content markers (see Bruns & Burgess, Chapter 1, this volume), Halavais (2013) suggests that 'they have been used as prompts for conversation, to crowdsource ideas or resources, and often to express sarcasm or parenthetical commentary on a tweet' (p. 37). Tsur and Rappoport (2012) also note that 'the use of hashtags is a popular way to give the context of a tweet, an important function due to the length constraint' (p. 645). Bruns and Stieglitz (2013) reinforce this notion of the hashtag as context by stating: 'A tweet consists of much more than just 140 characters' (p. 69). Both topical and nontopical hashtags can also be used to flag community affiliation or the desire to belong (Yang, Sun, Zhang, & Mei, 2012). Yang et al. liken the hashtag to a 'coat of [arms]' that can be flaunted to demonstrate community membership. Weller, Dröge and Puschmann (2011, p. 2) suggest that hashtags support the 'spontaneous creation of networks based on shared interests'. Rzeszotarski, Spiro, Matias, Monroy-Hernández and Morris (2014) refer specifically to 'Q&A hashtags' which users employ to post questions and seek advice or information. They suggest that these types of hashtags function 'both as a topical signifier (this tweet needs an answer!) and to reach out to those beyond their immediate followers (a community of helpful tweeters who monitor the hashtag)'.

Clearly, hashtags coordinate conversations, provide context and enable people to participate in discussion and request information. They can be emotive or topical, long-lasting or ephemeral, user-generated or prescribed. In this chapter, we use a case study of the #agchatoz hashtag to take into account how the hashtag not only serves to coordinate discussions and their participants but also plays a role in shaping these discussions themselves. In this way, we consider the *performative* role of the hashtag in materially shaping and coordinating public communication on specific issues, within and across social media platforms, as well as the performativity of the available methods for studying them. Because hashtags have material as well as symbolic features, they do not merely coordinate but also shape the dynamics of the 'ad hoc publics' (Bruns & Burgess, Chapter 1, this volume) that can be activated around an issue—indeed, we might even speak of how specific hashtags call specific publics into being.

This chapter focuses specifically on how some (but not all) hashtags can be understood as what Michel Callon and colleagues, in the context of technology and society studies, have called 'hybrid forums':

> Forums because they are open spaces where groups can come together to discuss technical options involving the collective, hybrid because the groups involved and the spokespersons claiming to represent them are heterogeneous, including experts, politicians, technicians, and laypersons who consider themselves involved. They are also hybrid because the questions and problems taken up are addressed at different levels in a variety of domains. (Callon et al., p. 18)

This conceptual framework is useful in our efforts to use social science research methods to understand emergent issues marked by complexity and uncertainty, because such an approach requires that we pay empirical attention to multiple, diverse perspectives—not only those of the most visible stakeholders but also those of affected constituencies who may not explicitly see themselves as stakeholders and/or may engage in non-normative, playful or 'unhelpful' ways (Michael, 2012). In exploring the possible forms and forums of 'technical democracy', e.g., in relation to nuclear power or genetically modified food, Callon et al. are discussing rather more formalised and more recognisably institutional spaces—indeed, the traditional institutions and fora of democracy. But in the contemporary media environment, digital methods have much to offer—not only because social media data can be used to identify and understand the dynamics of these hybrid forums but also because social media platforms are in themselves platforms for the emergence and coordination of hybrid forums.

Further, social media is particularly relevant to the idea of the hybrid forum because of the convergence of everyday, interpersonal and public communication that is so deeply constitutive of all social media platforms—whether in Twitter, Facebook, Instagram, or Tumblr, we can observe official government information, live documentation of street protests, memes drawing on popular culture, and everyday snapshots playing an equal part in the public communication around a controversial topic; and hashtags are very often used to introduce such content objects into the stream of an issue-related public. And adding another layer of hybridity, the traces and dynamics of such contested issues emerge within a complex and hybrid media environment (Chadwick, 2013) partly constituted via social media *platforms*, whose volatile dynamics, material features and competing business models also need to be taken into account (Burgess, 2014).

THE #AGCHATOZ CASE STUDY

Overview

A local variant on the original U.S.-based **#agchat** farmer advocacy or 'agvocacy' Twitter community, the AgChat Oz 'digital online community' co-founded by Tom Whitty, Danica Leys and Sam Livingstone originally had a mission to 'raise the profile of Australian agriculture by shining a light on the leading issues that affect the industry and the wider community'.[1] Weekly Twitter Q&A sessions use the **#agchatoz** hashtag to capture discussions of interest to the self-identifying agricultural community, ranging from personal issues such as succession planning and rural mental health to work matters including sustainable farming methods and how to manage natural disasters, as well as more public concerns such as animal welfare and live export. Most discussions solicit a range of perspectives from producers, consumers, scientists, journalists and other professionals; sometimes discussions connect to other issues and their hashtags (such as **#banlive-export** for the issue of animal welfare in the meat industry), thereby causing a collision of constituencies. Both the hashtag and its organisers have been covered a number of times by the ABC (Australia's primary public service broadcaster) and have close relationships with it, especially its rural affairs news division and programming.[2]

In its explicit self-framing as a forum, its persistent weekly 'meet-ups' and the way it is understood by its most regular participants as a meeting place for agriculturalists, **#agchatoz** operates more like a large, distributed and loosely organised focus group (Callon et al., 2009, pp. 164–165) than an 'ad hoc public' (cf. Bruns & Burgess, Chapter 1, this volume); as we will see, however, its existence in the dynamic environment of Twitter provokes encounters with a wider range of issues, bringing **#agchatoz**'s self-defined 'community' of users into contact with other constituencies and stakeholders. Particularly acute issues and controversies mediated via **#agchatoz** can be shown to quite significantly expand the 'inventories' (Callon et al., 2009, pp. 28–30) of both agricultural issues and issue publics.

In the study on which this chapter is based, we used a custom installation of the YourTwapperkeeper tool to collect 7 months' worth of data (from 19 July 2013–19 February 2014) comprising 73,218 tweets containing the **#agchatoz** hashtag. This dataset was subjected to a standard set of metrics (Bruns & Burgess, 2012; Burgess & Bruns, 2012) with the aid of Gawk and Tableau, including identifying the patterns of activity over time (including correlations between particular topics or issues and particular levels of volume and diversity in the participation), identifying the most visible and most resonant accounts (an inventory of actors), observing and visualising the network structure of the hashtag's participating accounts (an indication of the communities and/or publics associated with the hashtag) and exploring the co-occurrence of other hashtags with **#agchatoz** (an indication of hybridity and diversity of issues and publics). High-resolution versions of the data visualisations and some of the processed data sets are available via a Digital Appendix to this chapter.[3]

What Are the Patterns of Activity Over Time?

Weekly discussions form the foundation of **#agchatoz** hashtag use. Each month usually includes three topics set by the moderators and one 'mixed bag' session of questions on any topic. Participation is relatively stable, with notable spikes around particular topics, and a lull during the December–January seasonal hiatus. The greatest spike in activity during our sample was the 3 December 2013 'Future of **#agchatoz**' discussion. Questions revolved around desired values and outcomes, as well as how the community could ensure 'all members feel welcomed, respected and heard'. This discussion had been preceded by the fourth-largest spike of activity in our sample: the 19 November 2013 topic '**#AgChatOZ**—the role of new media in telling agriculture's story', with questions focussed more generally on how social media could be used to support and grow connections between producers and consumers. Both discussions included significant and sustained engagement around how best to deal with different cultural values and individual opinions, both within the core producer-based membership and with those outside the core group, including consumers and activists. The second-largest spike in activity from our sample was the 6 August 2013 topic 'Federal Election 2013', and the 3 September 2013 'Election Chat' follow-up was the fifth-largest spike. Questions from both discussions focussed on perceived differences between political rhetoric and actual policies or initiatives, asking about specific benefits the industry could expect from different parties. Both times, the question was posed whether or not rural Australia's expectations of government were too great. The third-largest spike in activity was the 1 October 2013 topic of animal welfare, although the related 22 October 2013 'Pest Animal Chat' did not garner especially great interest or engagement. The animal welfare chat also stood out for its explicit engagement guidelines:

> #AgChatOZ is Australia's leading platform to engage both rural and urban populations about agriculture and food. We are an open forum and invite all matters of opinion. Everyone is entitled to speak and share their view. Let's raise the standard tonight. This is not a platform for abuse or hatred, but a place that everyone is welcome for open dialogue. Through respectful conversation we can truly raise the bar.

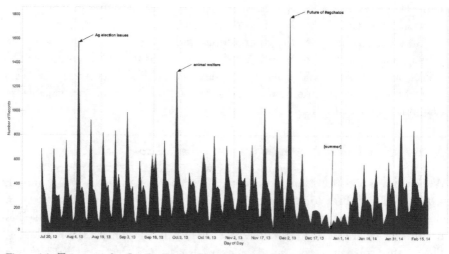

Figure 4.1. Tweets per day. See the Digital Appendix for high-resolution version.

But given our focus on 'hybrid forums', it is also important to note when there are spikes in the diversity of the discussions (more unique users per day = more diversity). The greatest diversity of membership in our sample can be seen in the 4 February 2014 topic 'Government assistance & relief packages in the Agricultural industry', and notable diversity is also seen in the 21 January 2014 disaster recovery discussion. The second-greatest diversity of membership can be seen in the 11 February 2014 topic 'Are we becoming Asia's Food Bowl?'—though this was not one of the liveliest discussions overall. Significant diversity is also seen in the 8 October 2013 topic 'Future Ag industries for Aus & the skills required to meet the demand...' and the 22 October 2013 'Pest Animal Chat'. Notably, although the pest animal chat had lower overall participation than the animal welfare chat, it had a slightly more diverse membership—the politics of human-animal relations energise debate across society. And finally, diversity of membership is also notable in the 14 January 2014 topic 'The Year Ahead'—perhaps reflecting a larger number than usual of participants 'checking in' after the traditional summer break over Christmas and New Year.

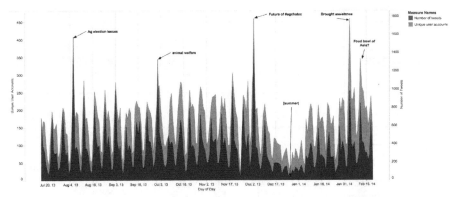

Figure 4.2. Tweets and unique accounts per day. See the Digital Appendix for high-resolution version.

Using Shared URLs to Inventory the Issues

As an initial probing exercise into the inventory of issues associated with **#agchatoz** beyond the nominated topics for each week's 'meetup', we extracted the most shared URLs over the 5 months of data collection. The aggregated result gives an indication of the kinds of topical coverage that most mobilise and engage the **#agchatoz** participants.[4] By proxy, it also provides an initial sense of the most resonant issues and perspectives on them, as well as indications of new or unexpected actors becoming associated with the hashtag.[5]

Among the 25 most-shared URLs are several images, including humorous or political images, mostly organised around pro-framing advocacy, as well as climate change and coal seam gas (CSG) debates. For example, one of the most shared images, captioned (via the originating tweet), 'how to win the climate change debate in one picture' is an infographic showing current and historical average temperatures around Australia; another promotes Australian Bacon Week and another is a Twitter pic of a farm sign reading as follows:

> Notice: This property is a farm. Farms have animals. Animals make: funny sounds, smell bad, and have sex outdoors. Unless you can tolerate: noise, odors, and outdoor sex, don't buy a property next to a farm!

There are also links to informational, political and advocacy resources, including a press release about a New South Wales (NSW) farmers workshop for landholders on mining and coal seam gas, and an article in *Scientific American*, 'Are Fracking Wastewater Wells Poisoning the Ground beneath Our Feet?', and a more pointed YouTube video shared by @QldCountryLife—a voice-over with stills presented by an Australian farmer addressed at urban greenies with a lack of respect for farmers, inviting them to 'get back to [him]' when they inevitably realise how much they need Australian farmers to survive. Also frequently shared was a press release by Australian National Party senator Barnaby Joyce urging us to 'keep a farmer in mind on World Mental Health Day' on 10 October 2013, as well as links to locavore blogs and campaigns against imported fruit and vegetables.

As we might expect, the issues most often circle around the most explicitly stated primary concerns of the core **#agchatoz** constituency. One of the primary activities represented in the **#agchatoz** conversation is 'agvocacy', which includes both organised lobbying and more ad hoc, vernacular and playful modes of expression; through to structured, deliberative democratic engagement with high-stakes environmental issues affecting farming and rural communities, such as climate change and coal seam gas exploration; and even at times creating some counterintuitive alliances between the urban left and the rural right—even bringing the **#agchatoz** community into an unlikely alliance with the Greens (cf. the anti-CSG/fracking Shut the Gate campaign).

While much of the tenor of the conversation frames the hashtag as an opportunity to bypass media stereotypes and have a voice in national debate, there is also a fair bit of antagonism towards a supposedly uninformed city-dwelling culture who insufficiently value the role of agribusiness in Australia's society and economy, and there are some dramatic collisions of opposing viewpoints and organised political groups on issues like animal welfare and animal rights. But **#agchatoz** and its counterparts in other countries are also part of the vernacular culture of the Web, illustrated most colourfully by the international phenomenon of 'felfies'—short for 'farm selfie', where farmers pose for self-portraits in situ on the farm, often with a favourite sheepdog, calf, or tractor as co-star.[6]

#agchatoz and Its Social Networks

A first step in understanding how the **#agchatoz** participant pool might relate to existing issue- or interest-based communities was to visualise the users in the dataset as a social network.[7] The map is based on a data snapshot of follower-followee relationships among all accounts in the overall dataset. The snapshot was taken in April 2014, so it doesn't represent the dynamic, fluctuating ways that more or less permanent relations might be established or dissolved by participants in the **#agchatoz** hashtag, but in some ways it could be viewed in cumulative terms, as the #agchatoz meetups had been going on for a couple of years by this point (see Figure 4.3, and the Digital Appendix for a high-resolution version).

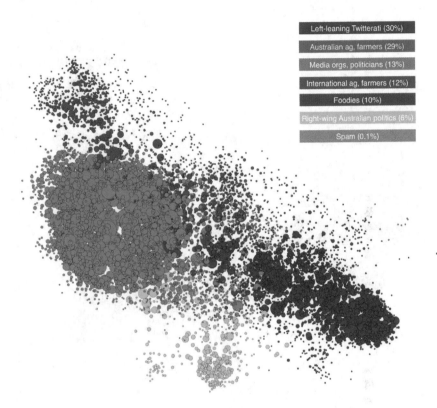

Left-leaning Twitterati (30%)

Australian ag, farmers (29%)

Media orgs, politicians (13%)

International ag, farmers (12%)

Foodies (10%)

Right-wing Australian politics (6%)

Spam (0.1%)

Figure 4.3. Follower-followee network, nodes sized according to number of #agchatoz followers. See the Digital Appendix for a high-resolution version.

There is little contrast between the first two maps, indicating a very high degree of mutual follower-followee relationships, which we might take as indication of a strong core community of regular participants. There are quite distinct (perhaps even unusually distinct) 'communities' identified by the algorithm, which with manual coding have been identified as organised around Australian agricultural and farming organisations (the largest cluster), neighbouring the international agricultural community, with other clusters largely populated by 'foodies' (including urban food producers or retailers, cooks and locavore gourmets), traditional media organisations and MPs (especially rural ABC programming or journalists and rural politicians), and then the 'general' Twitter communities structured around progressive politics and digital culture and conservative Australian politics, respectively. The very co-presence of such apparently different social groupings within a hashtag discussion is already generative of political hybridity (see Figure 4.4, and the Digital Appendix for a high-resolution version).

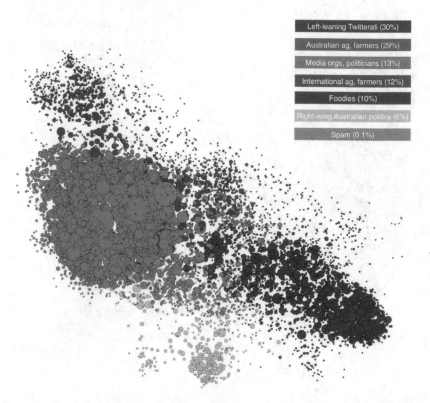

Left-leaning Twitterati (30%)

Australian ag, farmers (29%)

Media orgs, politicians (13%)

International ag, farmers (12%)

Foodies (10%)

Right-wing Australian politics (6%)

Spam (0.1%)

Figure 4.4. Follower-followee network, nodes sized according to number of #agchatoz followees. See the Digital Appendix for a high-resolution version.

However, the third and fourth maps—in which nodes are sized according to their total Twitter followers overall and their level of tweeting activity on the **#agchatoz** hashtag, respectively—highlight some of the distinctive features of a deliberately constituted 'hybrid forum' such as this (see Figures 4.5 and 4.6, and the Digital Appendix for high-resolution versions).

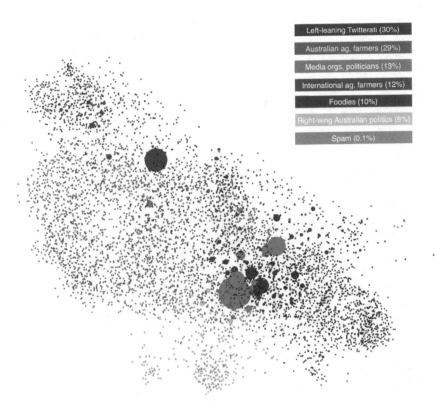

Left-leaning Twitterati (30%)

Australian ag, farmers (29%)

Media orgs, politicians (13%)

International ag, farmers (12%)

Foodies (10%)

Right-wing Australian politics (6%)

Spam (0.1%)

Figure 4.5. Follower-followee network, nodes sized according to overall number of followers. See the Digital Appendix for a high-resolution version.

This helps us to see that there is a core of highly vocal users on the hashtag—the original organisers, the @agchatoz account itself, and a small number of highly active agricultural stakeholders. And most of these accounts have been far more active in the **#agchatoz** data than they have been overall over the life of Twitter as a platform, or they are relatively young accounts in comparison to the 'Twitterati' group on the lower right-hand side, or both (see Figure 4.7, and the Digital Appendix for a high-resolution version).

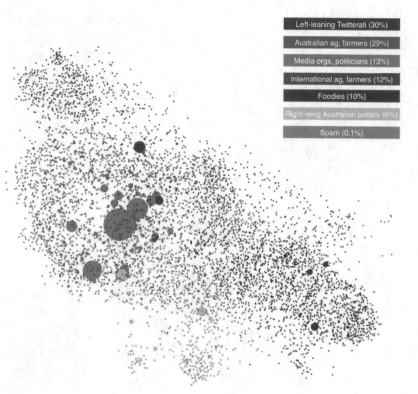

Left-leaning Twitterati (30%)
Australian ag, farmers (29%)
Media orgs, politicians (13%)
International ag, farmers (12%)
Foodies (10%)
Right-wing Australian politics (6%)
Spam (0.1%)

Figure 4.6. Follower-followee network, nodes sized according to number of #agchatoz tweets. See the Digital Appendix for a high-resolution version.

Left-leaning Twitterati (30%)

Australian ag, farmers (29%)

Media orgs, politicians (13%)

International ag, farmers (12%)

Foodies (10%)

Right-wing Australian politics (6%)

Spam (0.1%)

Figure 4.7. Follower-followee network, nodes sized according to total all-time tweets. See the Digital Appendix for a high-resolution version.

Using Co-hashtag Analysis to Inventory Issue Publics

Generating lists of the other hashtags that co-occur with **#agchatoz** in the data helps us to begin to understand the relations between **#agchatoz** topics and some of the many other issues, networks and communities that are coordinated via Twitter hashtags. A total of 5,903 additional hashtags appear in the data set.[8] The most commonly used hashtag within **#agchatoz** tweets is **#auspol** (as the primary hashtag for discussion of Australian politics, one of the Australian Twittersphere's favourite topics, it unsurprisingly occurred 3,750 times in the dataset), followed—often in the same tweet—by **#csg** (coal seam gas, 2,461 occurrences) and **#nswpol** (New South Wales politics, 1,382 occurrences). Rounding out the top five additional hashtags are **#freshproduce** (1,382) and **#pmaanznews** (1,377), both related to the Produce Marketing Association Australia–New Zealand. In addition to other state-based politics (**#qldpol**) and sectoral structural issues and events (**#drought, #nswfires, #harvest13**), individual agricultural industries are also well represented

(#ausdairy, #beef, #wheat), as are cross-links to other #agchat communities in New Zealand, the U.K. and the U.S. (#agchatnz, #agchatuk, #agchat). The most common issues tied to the #agchatoz hashtag are specifically farmer based, such as drought or youth in agriculture; more broadly and publically contentious, such as mining and live export; or, less commonly, climate change and genetically modified organisms (GMOs). These topics involve the widest range of publics coming into contact with each other around shared issues of concern but not necessarily shared politics (#fracking, #gas, and #frackoff; #supportlivex, #animalwelfare and #banliveexport; #woolworths, #whycoles). Further down the long tail, the more mundane, human and affective aspects of rural life—with significant potential to engage the wider community—are present (#strongwomen, #farmpics, #felfie, #morningwalk, #youngfarmer, #passion).

CONCLUSIONS

Following Callon et al. (2009), we argue that the sub-issues and controversies that play out via #agchatoz as a hybrid forum have the potential to act as 'apparatuses for exploration and learning' (p. 35). A combined reading of the #agchatoz network structures and the co-hashtag analysis provides preliminary indications that the collision of different issues and their publics in the social media environment may lead to the 'discovery of mutual, developing, and malleable identities that are led to take each other into account and thereby transform themselves' (p. 35) and that in part, this potential is provoked by the ways in which hybrid forums trouble the 'two great divisions' of Western societies: the distinction between domain specialists and laypersons and the distinction between ordinary citizens and their political representatives (Callon et al., 2009, p. 35). Working within this conceptual framework, we have demonstrated how digital methods might be used to go beyond noting the loudest voices and dominant themes in binaristic debates and instead trace more of the diversity of stakeholder and non-stakeholder perspectives, the substantive issues and topical diversions that come together within the kinds of hashtags that appear to operate as hybrid forums.

The communities and issues most obviously associated with #agchatoz are quite predictable in the context of Australian politics on Twitter. But down the 'long tail' the topical coverage becomes more diverse, less predictable, but also highly evocative of adjacent issues and publics where there might be considerable potential—if not yet actualized—for #agchatoz to provoke a redefinition of what counts as 'agricultural' politics in Australia and who counts as a stakeholder in such a politics. There might also be potential to actually promote a more enriched, dialogical model of democratic engagement, in the context of a hybrid forum with the participation of not only farmers and agricultural organisations but also the national and local media, rural and urban food producers (and their animals!), politically and environmentally engaged consumers and politicians at all levels of government. Future research and practical experimentation might explore the possibilities to highlight, support and amplify the hybridity of forums like **#agchatoz** and mobilise it in the service of more dialogical modes of political engagement.

NOTES

1. See http://www.agchatoz.org.au/what-is-agchatoz
2. See, for example, http://www.abc.net.au/news/2013-05-30/the-bush-bytes-back/4722248
3. The Digital Appendix is at http://mappingonlinepublics.net/?p=2998
4. The complete list of the most-shared URLs is available as a Google Spreadsheet via the Digital Appendix.
5. This list is based on the occurrence of shortened URLs only (using the Twitter t.co format), not expanded ones. Therefore we are likely to be over-representing retweets over original mentions. We have expanded these shortened URLs for ease of understanding here; broken and incomplete links were removed to produce a top 50 from 57 containing 7 incomplete links.
6. See, for example, Gunders (2014).
7. Follower-followee network graphs were produced by querying the Twitter API for the follower and friends lists for each username in the **#agchatoz** dataset as at April 2014 and then constructing a network graph of follower-followee relations among **#agchatoz** participants in Gephi, adding overall status counts, overall followers, and number of tweets in the **#agchatoz** dataset as supplementary node attributes. The network was spatialised using the Force Atlas 2 algorithm. The visualisation was then filtered so that only nodes with a degree of 7 (a combined total of 7 follower or followee connections) remained visible. Gephi's modularity algorithm was used to calculate and then colour 'communities' whose members have higher-than-random affinity, resulting in a total of six visible clusters and a modularity score of 3.9 at the default resolution setting of 1.0. Manual review of the profiles attached to the Twitter accounts that clustered together in these 'communities' was used to identify some common characteristics; lists of participant account IDs were subjected to an additional, independent manual coding exercise to test the applicability of the resulting labels.
8. The full list of additional hashtags is available as a Google Spreadsheet via the Digital Appendix.

REFERENCES

Bruns, A., & Burgess, J. (2012). Researching news discussion on Twitter: New methodologies. *Journalism Studies, 13*(5–6), 801–814.

Bruns, A., & Moe, H. (2013). Structural layers of communication on Twitter. In K. Weller, A. Bruns, J. Burgess, M. Mahrt, & C. Puschmann (Eds.), *Twitter and society* (pp. 15–28). New York: Peter Lang.

Bruns, A., & Stieglitz, S. (2012). Quantitative approaches to comparing communication patterns on Twitter. *Journal of Technology in Human Services, 30*(3–4), 160–185.

Bruns, A., & Stieglitz, S. (2014). Metrics for understanding communication on Twitter. In K. Weller, A. Bruns, J. Burgess, M. Mahrt, & C. Puschmann (Eds.), *Twitter and society* (pp. 69–82). New York: Peter Lang.

Burgess, J. (2014, January 16). From 'broadcast yourself' to 'follow your interests': Making over social media. *International Journal of Cultural Studies* [online first]. Retrieved from http://ics.sagepub.com/content/early/2014/01/13/1367877913513684.abstract

Burgess, J., & Bruns, A. (2012). Twitter archives and the challenges of 'Big Social Data' for media and communication research. *M/C Journal, 15*(5). Retrieved from http://www.journal.media-culture.org.au/index.php/mcjournal/article/viewArticle/561

Callon, M., Lascoumes, P., & Barthe, Y. (2009). *Acting in an uncertain world. An essay on technical democracy* (G. Burchell, Trans.). Cambridge, MA: MIT Press.

Chadwick, A. (2013). *The hybrid media system: Politics and power.* Oxford, UK: Oxford University Press.

Deller, R. (2011). Twittering on: Audience research and participation using Twitter. *Participations, 8*(1). Retrieved from http://www.participations.org/Volume%208/Issue%201/deller.htm

Gunders, P. (2014, January 15). 'Herd' of a felfie? ABC Southern Queensland. Retrieved from http://www.abc.net.au/local/photos/2014/01/15/3926243.htm

Halavais, A. (2013). Structure of Twitter: Social and technical. In K. Weller, A. Bruns, J. Burgess, M. Mahrt, & C. Puschmann (Eds.), *Twitter and society* (pp. 29–42). New York: Peter Lang.

Marres, N. (2012). *Material participation: Technology, the environment and everyday publics.* London: Palgrave.

Michael, M. (2012). What are we busy doing?: Engaging the idiot. *Science, Technology & Human Values, 36*(5).

Papacharissi, Z. (2010). *A private sphere: Democracy in a digital age.* Cambridge: Polity Press.

Rogers, R. (2013) *Digital methods.* Cambridge, MA: MIT Press.

Ross, C., Terras, M., Warwick, C., & Welsh, A. (2011). Enabled backchannel: Conference Twitter use by digital humanists. *Journal of Documentation, 67*(2), 214–237.

Rzeszotarski, J. M., Spiro, E. S., Matias, J. N., Monroy-Hernández, A., & Morris, M. R. (2014). *Is anyone out there? Unpacking Q&A hashtags on Twitter.* Paper presented at the ACM CHI Conference on Human Factors in Computing Systems, April 26–May 1, Toronto.

Tsur, O., & Rappoport, A. (2012). What's in a hashtag?: Content based prediction of the spread of ideas in microblogging communities. In *Proceedings of the 5th ACM international conference on Web search and data mining* (pp. 643–652). Seattle: ACM.

Weller, K., Bruns, A., Burgess, J., Puschmann, C., & Mahrt, M. (2014). *Twitter and society.* New York: Peter Lang.

Weller, K., Dröge, E., & Puschmann, C. (2011). *Citation analysis in Twitter: Approaches for defining and measuring information flows within tweets during scientific conferences.* Paper presented at Making Sense of Microposts Workshop (# MSM2011). Co-located with Extended Semantic Web Conference, Crete. Retrieved from http://ceur-ws.org/Vol-718/paper_04.pdf

Yang, L., Sun, T., Zhang, M., & Mei, Q. (2012). We know what@ you# tag: Does the dual role affect hashtag adoption? In *Proceedings of the 21st international conference on World Wide Web* (pp. 261–270). New York: ACM.

CHAPTER FIVE

#Time

DANIEL FALTESEK

Perhaps the most important distinction between Twitter and other social network sites has been the claim that Twitter offers access to real time. But what is real time? There is nothing real about the construction of time on Twitter. As time seemingly moves forward, old tweets are displaced by new tweets at the top of the screen. Depending on how many people one follows, time passes at dramatically different rates. To say something is trending now is a misnomer: there is no unified now, only a collection of page alignments linked by gossamer tissues of calls and writes to an SQL database. Acronyms are the building blocks of the social now, and that now is created only when hailed into existence by a call to the database. If anything, the procession of signifiers on Twitter and through a hashtag public is neither real nor time, which is not to say that it is unreal or unimportant, but that the investment of a machine process with the creation of real time is a powerful affective move.

Analysis of the formation of publics at the level of the tweet is difficult if not impossible because of this tenuous machine logic. The relations between these writers and readers are distinguished not spatially but temporally. Aside from the isolation of locally trending tweets (this can also be changed), Twitter is spatially ambivalent. After a short time, tweets are no longer accessible. Tweets vanish. Time is everything on Twitter. Products such as Buffer, HootSuite, and Tweet-deck promise tweeters the ability to plan their tweets in advance and to manage multiple flows of activity. Nothing is quite as important for marketers as building an effective sense of cadence, expressed as a particular rhythm for inserting messages into other people's social media lives. Cadence theory offers strategic communicators the promise of transcending the momentary bounds of Twitter by preparing well in advance content that might be automatically circulated to users during high-traffic times. The dream of strategy is to escape from the bounds of conversation and intersubjective engagement to manipulate the stream of time. The great selling point of Twitter—real time—is intended only for persons who are consumers, not the power users of enhanced versions of the platform.

Fan-out, the process by which tweets are populated into readers' feeds, draws attention to the problem of time in a very concrete way, as tweets can arrive out of order (Hoff, 2014). This tends to happen when users have asymmetric follower lists, or just through the random chance of location and web traffic. For all the emphasis on the production of a stable timeline and a real-interactive moment, even the seemingly sacrosanct flow of real time can be out of order.

Temporality (the phenomenological experience of time) is distinct from an attention to history, or to time itself. Or to use rhetorical terminology, *kairos* (the rhetorical moment or situation) is distinct from *chronos* (the sequence of events). *Kairos* cannot be constituted on the level of the individual, unread tweet; it is, after all, a production of the attention of at least several people and mechanical processes. Although one might read a stream of tweets and recognize their inter-connectivity, this sort of deep reading would require an attention to contextual cues that may simply be unavailable in the stream. Users often use a combination of addressivity through the @ and .@ signs as well as the hashtag to provide a lateral link between signs. Overt signifiers make the connections visible with even fleeting attention. Our concern should not be for the properties of the lateral links themselves or even for the content of the communications but for the contexts created and destroyed through the temporality of Twitter. In this chapter, I argue that the temporal features of networks organized by hashtags call for a particular attention to the time in which networks operate and, further, that the temporality of circulation of discourses on these networks is presented both as a form of fleeting publicity as well as a form of simulated publicness. This essay is concerned foremost with a consideration of the when of a public rather than the why, how, or where.

TIME- AND SPACE-BINDING MEDIA

Harold Innis's distinction between time-binding and space-binding media has informed a great deal of work in media studies, particularly James Carey's *Communication and Culture* (1989). Time-binding media persist across time, carrying messages to future generations, such as a monument. Space-binding media overcome the limits of place, reaching many people at the same time, such as radio. These conceptions of media and culture tend to trade off; cultures that privilege space-binding media become increasingly oriented toward the present, and those that emphasize continuity across time are tied to a particular place (Innis, 1950, p. 7). These biases are dialectical, as time-binding media require a form of storage to transmit to the future, and space-binding media tell stories that exist in a particular time.

Social media technologies seem to be space binding: they transmit to large publics quickly and are seemingly impermanent. A tweet seems more like a radio broadcast than a monument. As it has become clear that the memory of the Internet is functionally unlimited, the time-binding character of social networks has become a point of anxiety. Unlike other space-binding media, the default for social media items is long-term storage. Anxiety about social media is understandable, as these technologies skew established categories. This takes a number of forms, perhaps most curiously in calls for forgetting.

Snapchat's rapid rise in popularity attests to the idea that space-binding affordances should be kept separate from time-binding affordances. Or, the kind of communication that a platform facilitates should be clear. Snapchat restores this balance through the logistical marker of the ticking clock—once opened, a snap exists for the next 10 seconds before permanent deletion, or at least that is the firm's claim. Visible countdown technologies are a powerful cultural resource marking the anterior of a future event, be that the erasure of an image, the destruction of the *Enterprise* and the victory of James T. Kirk, or the conclusion of a sporting event. Running out of time, and the cultural practice of watching the clock, become performative enactments of preservation and the shared experience of the moment.

Timekeeping devices are, as John Durham Peters writes, a form of logistical media—they organize the distribution of people and things, they are the media that organize all others (2009, p. 16). The experience of the flow of time must be considered in the history of logistical media but also through the interplay of those mechanisms with linguistic codes and phenomenological experiences. Returning to the question of the introduction, it is important to note that absent attention, timekeeping technologies have little impact on a society. In *A Plea for Time*, Innis cautions media scholars not to overplay the idea of the moment in all its romance:

> We must somehow escape on the one hand from our obsession with the moment and on the other hand from our obsession with history. In freeing ourselves from time and attempting a balance between the demands of time and space we can develop conditions favourable to an interest in cultural activity. (1950, p. 21)

An attention to time is thus distinct from an attention to either history, by stabilizing an archive of things that might be cataloged, or the moment itself—at least as that moment is understood in an exclusively phenomenological register. The study of time and of the logistical media that organize publics should be distinct from either the study of history or the feeling of the moment.

John Durham Peters noted that communication scholarship has also considered time as a point, or *kairos* (2009). This view has informed work on the side of rhetorical studies through the question of the rhetorical situation. As a theory of time, it appears far less frequently than a discussion of context, creativity, or materiality. This debate offered a way of viewing the seemingly objective procession of the historical moment for Lloyd Bitzer (1968). Speakers and publics formation was facilitated by the procession of historical and material causes. Instead of speakers imaging situations to respond to, the exigency exists before the speaker. Richard Vatz responded to this in 1973, arguing that speakers may produce the situation to which they respond. One can argue about the benefits of war with a country without actually resting a finger on a missile launch control. This is not to say that all rhetorical situations offer equal possibilities for persuasion or the legitimation of a policy proposal.

Shifting this debate, Barbara Biesecker's (1989) turn toward Derridian deconstruction offered a way to see the positions as dialectically tied aspects of the same metaphysical sense of time. Instead of seeing the production of the rhetorical situation as being related to the character of the moment or of the speaker, Biesecker argued that the persuasive moment exists because of the potential of language itself. This use of Derrida offers the prospect of finding the identity and temporality of the public and the rhetorical situation not in the act of making a single hashtag itself but in the possibility of attention to the moment of being with a hashtag. For William Trapani, Biesecker's use of Derrida is an important development in the argumentative understanding of the moment, an opportunity to introduce Derrida's later work on media technologies that produce these new forms of meaning at work in the flow of the online rhetorical situation in the interplay between multiple, divergent, out-of-phase temporalities. Trapani argues that Derrida's conception of the event-machine offers three key ideas: the iterability of speech acts, their dissemination in unexpected ways between mechanical and performative processes, and the plasticity of temporality (2010, p. 329). As a description of the ongoing process of public formation and articulation through social network systems, this is particularly apt. A like or a retweet is nothing if not a speech act that, depending on the peculiarities of the interaction of many servers and other pools of attention, might flow unpredictably from person to person and feed to feed.

This extension of the deconstruction of the rhetorical situation aptly describes how the formation of publics through hashtags as the condition of their existence depends on the possibility of différance existent in the iteration, citation, and circulation of discourses at simultaneous times, yet through divergent, even singular, experiences. Derrida makes this argument well in the context of the temporal dimension of e-mail:

> But the example of E-mail is privileged in my opinion for a more important and obvious reason: because electronic mail today, even more than the fax, is on the way to transforming the entire public and private space of humanity, and first of all the limit between the private, the secret (public or private), and the public or the phenomenal. It is not only a technique, in the ordinary and limited sense of the term: at an unprecedented rhythm, in quasi-instantaneous fashion, this instrumental possibility of production, of printing, of conservation, and of destruction of the archive must inevitably be accompanied by juridical and thus political transformations. (1998, p. 17)

In the context of Derrida's study of the archive and the meaning of the event, the interplay between the technical capacity of the inscription system and the meaning of the event come full circle. The connection of these two distinct accounts of the production of argumentative time can be productive as it emphasizes the slippage between media forms and the composition of publics. Logistical media theories provide the basis for understanding the literal physical timekeeping machines that orchestrate social networks. The deconstructed rhetorical situation provides insights into the texts coordinated by those machines. For some, these conceptions of an attention to logistical media and the event-machine may be mutually exclusive; I would contend that they are both necessary for understanding the ways in which uneven, seemingly random space-binding network machines interact with the affective potential of different, laterally linked, hashtagged publics. The idea of phase can be particularly useful for understanding the relationship between different forms of power. It is this attention to both the logistical means by which social network publics are coordinated and the potential of their textual unfolding that might offer a contribution to understanding the operation of publics through social networks. In the next sections, I will call attention to two important ideas in public sphere theory that call for an understanding of time.

FLEETING PUBLICS

The idea of a fleeting or momentary communication being necessary for suppression or for promotion is really quite common. For example, fleeting expletives may allow a broadcaster to avoid a fine while an apparently deliberate obscene act draws sanctions, if only to enforce a certain set of notions about time on the public airwaves.

Michael Warner argues that participation in the life of a public or counterpublic need not include address and response: "mere attention" is enough to constitute engagement. This is profoundly important for understanding what it means to constitute a public on Twitter. The idea is not that everyone is engaged and responding; this would become a sea of incoherent meaning, and frequently is. Often, these relationships exist by chance:

> You might be multi-tasking at the computer; the television might be on while you are vacuuming the carpet; or you might have wandered into hearing range of the speaker's podium in a convention hall only because it was on your way to the bathroom. (Warner, 2002, p. 419)

Instead of offering a permanent theory of the public, Warner's publics exist in moments that are all too often unstable or intermittent. This distinction is important, as the institutional logic of system-oriented publics depends on a strict regulation of who may have access and how they may behave. If publics form by mere attention, their memberships are fluid and provisional, their boundaries are porous.

For Warner, publics form only inasmuch as they are addressed by discourse in circulation. A public is not a historical form but a literal, present-tense thing that is created through attention. It is this condition of publicness that allows the formation of accidental publics and covert publics. If we take as a guide those who argue that "public sphere" is a poor translation of *Öffentlichkeit*, it is important to recognize that this form of ongoing publicness in a situated moment is what Habermas was attempting to describe: publicity as a process is not dependent on spaces but on the time of attention and readiness. It is surely possible that not all people addressed by the discourse of a public would become active recirculators of it. Not all tweets are retweeted. Many are never seen. This would lend to the impression of the feed as nonsense, as well as making the platform commercially useless. Without some kind of distinction between valuable, recirculable information and the information that will pass without further activation, it would be difficult to sell an advertising product.

Twitter was made for Warner's publics. Hashtag publics exist only because of the possibility of attention and recirculation. They, as Warner describes, "crave attention like a child" and yet can be formed by people who merely "wandered by" (2002, p. 419). Instead of provisional publics formed in kinky clubs or street-corner ministries, the time of the public is found in that curious relation of the time of the server and the time of the profile, made stable by the echoes of retweets and the time of those who read the tweets—or at least those who are ready to understand what they might mean.

SIMULATED PUBLICNESS

Habermas noted in *Structural Transformation of the Public Sphere* (1989) that artificial publicities could just as easily use the techniques of publicity for purposes other than the actualization of political change (p. 247). The tendency for the capacity of the system's logic—be that the public system, the government; or the private system, the market—is quite real. At the time his book was originally published, Habermas was sanguine about the future of the publicity process:

> The immediate effect of publicity is not exhausted by the decommercialized wooing effect of an aura of good will that produces a readiness to assent. Beyond influencing consumer decisions this publicity is now also useful for exerting political pressure because it mobilizes a potential of inarticulate readiness to assent, that, if need be, can be translated into a ple-biscitarily defined acclamation. (1989, p. 201)

As a preemptive description of the interaction of "inarticulate readiness" with institutions of power, Habermas dreams Twitter into existence. The key idea here is that the possibility of engaging a public of a rapidly refreshing social network is contingent on the idea of the public being in a position to recognize that they are in fact in public. Advertising and other processes for cultivating attention are greeted with suspicion, as the aura they produce is often strategically designed to depend on less-than-rational and -critical deliberation. In many cases, the sorts of publics that advertisers and public relations professionals wish to cultivate are synthetic. In an attempt to harness the political will of activist publics, Pepsico curtailed many its advertising efforts, redirecting money into various grants for good causes (Bida, 2012). Showering money on good people was supposed to engender goodwill by articulating the product, Pepsi, to a more robust sense of civic engagement. By the end of the campaign, Pepsi had fallen to third in the soft drink market.

Synthetic publics are not particularly satisfying. Although verification is difficult, the swell of followers of Newt Gingrich's Twitter account during the 2012 Republican primary process is an important example (Cook, 2011). Those followers were likely sock puppets, or fictional accounts used to manufacture the illusion of support or popularity. Gingrich's sudden surge in followers was not intended to persuade those followers but to create the conditions where by the locus of his campaign energy could be valued, to create the logic of intelligibility that would say that it was a Twitter handle worth following.

Paul Virilio's (2003) theory of acceleration offers an important insight into the diffusion of information through Twitter. Instead of the millisecond culture of Twitter facilitating even faster processes of dialectical judgment and resolution, these same techniques and velocities magnify the accidents and distortions of the publicity process. False information and well-positioned conspiracy theories find increased reach through social media. The accident of the simulated public is the misrecognition of the nature of the formation of publics. One who is purchasing sock puppets misapprehends that some affinity, some desire, seeds the hubs of a network. Like a rose, a social network needs a trellis to climb. It is not the raw number of followers that make the celebrity, but one cannot be a celebrity without followers; attention at the right moment is the sine qua non of Twitter, and that moment exists in a fraught space between subjectivity and mechanical processes. Copying a follower list or purchasing fake accounts takes a decidedly historical approach to the formation of the public—as if a public could be constructed or simulated rather than formed as a result of attention. The connections and networks that a user might have cannot be read outside of the time in which they were made. Yet the public formed only because of a simulation of time in the first place. In this way, the creation of a synthetic world of public impression, publics forming on Twitter and their archives are of a curious character that moves beyond e-mail and other technologies. They call the basic idea of the impression, the public, and the archive to the fore to be read through time.

THEORIZING THE TIME OF THE #PUBLIC

The time of Twitter is measured in gigahertz. This is not the time of the hashtag public. Publics exist in the temporality of their circulation, and in the case of a system that auto-poetically produces circulations, those times will always be uneven. For some users, the flow of status updates from a large number of people and machines will make it almost impossible to read or listen to incoming messages. Meaning will wash away because it comes in too much and too fast. This chapter has argued that the stability across time provided by hashtags is an artifact of those that trend and that the experience of a laterally organized temporal social network is an operation of an event-machine that produces opportunities for political engagement and rhetorical critique both through the possibility of difference inherent in communication and through the differential action of server mechanisms.

Hashtag publics offer a chance for fleeting connections to become forms of attention that constitute publics. Yet the technologies that might afford publics the chance to form are also all too easily appropriated for forms of system logic. The temporalities of publics to be formed are synthesized through cadence software, sock puppets, and trends. Entire pedagogies are devoted to the manipulation of the time of Twitter life.

Studies of the circulation of hashtags need to pay particular attention to the temporality of their circulation as part of an unpredictable flow of messages that is both tightly controlled and beyond control at the same time. This practice of reading, thinking, and engaging must take place in the real time of the researcher inasmuch as the traces left by the hashtag are a poor substitute for the phenomena itself. The practice of data management at Twitter underscores this reality. The firehose, the raw stream of data flowing through the Twitter system, is carefully controlled—very few parties have access to the firehose, much less the API storehouse of old Twitter data. With good reason: that database and the operations and transformations to be done to it are the heart of Twitter's business plan. Further, providing unlimited search access to all of that data would be cost and technology prohibitive. Twitter is in this sense very human. It is a machine with a memory, and limits. As an event-machine, the hashtag public functions in the moment of its circulation; unless actions are taken at the time of circulation, tweets are lost to the ether of cyberspace. Even when archives are made of the hashtag, the rhetorical force of that moment is lost. In that moment of circulation and iteration between data handlers, people, and text, there is a possibility for persuasion and public formation that will not come again in a singular sense but very likely will come again in a different form when attention refocuses at some other point in time.

#Time requires an attention and a form of research that is situated in the feeling of that moment, practiced within the time of the hashtag, and committed to the difference of that very second. This calls us back to Innis; researching the hashtag public requires an attention both to the moment of possibility in which one tweets and to the history of the mechanism, if only because this attention might create a moment for a consideration of the time of the hashtag public itself.

REFERENCES

Bida, C. (2012, October 29). *Why Pepsi canned the Refresh Project*. Retrieved from http://www.media post.com/publications/article/186127/why-pepsi-canned-the-refresh-project.html

Biesecker, B. (1989). Rethinking the rhetorical situation from within the thematic of différance. *Philosophy and Rhetoric, 22*(2), 110–130.

Bitzer, L. (1968). The rhetorical situation. *Philosophy and Rhetoric, 2*(1), 1–14.

Carey, J. (1989). *Communication as culture*. New York: Routledge.

Cook, J. (2011, August 2). *Update: Only 92% of Newt Gingrich's Twitter followers are fake*. Retrieved from http://gawker.com/5826960/update-only-92-of-newt-gingrichs-twitter-followers-are-fake

Derrida, J. (1998). *Archive fever: A Freudian impression* (E. Prenowitz, Trans.). Chicago: University of Chicago Press.

Habermas, J. (1989). *The structural transformation of the public sphere: An inquiry into a category of bourgeois society* (T. McCarthy, Trans.). Cambridge, MA: MIT Press.

Hoff, T. (2013, July 8). *The architecture Twitter uses to deal with 150M active users, 300K QPS, a 22 MB/S firehose, and send tweets in under 5 seconds* [Blog post]. Retrieved from http://highscal ability.com/blog/2013/7/8/the-architecture-twitter-uses-to-deal-with-150m-active-users.html

Innis, H. (1950). *A plea for time*. Lecture delivered the University of New Brunswick, March 30, 1950. http://www.gutenberg.ca/ebooks/innis-plea/innis-plea-00-h.html

Peters, J. D. (2009). *Calendar, clock, tower*. Paper presented at the Media in Transition 6, April 24–26, Massachusetts Institute of Technology. Retrieved from http://web.mit.edu/comm-forum/mit6/ papers/peters.pdf

Trapani, W. (2010). Materiality's time: Rethinking the event from the Derridian espirit d' a-propos. In B. Biesecker & J. Lucaties (Eds.), *Rhetoric, materiality, and politics* (pp. 321–346). New York: Peter Lang.

Vatz, R. (1973). The myth of the rhetorical situation. *Philosophy and Rhetoric, 6*(3), 154–161.

Virilio, P. (2003). *Unknown quantity*. London: Thames & Hudson.

Warner, M. (2002). Publics and counter-publics. *Quarterly Journal of Speech, 88*, 413–425.

Hashtags and Activist Publics

Come Together, Right Now: Retweeting in the Social Model of Protest Mobilization

AARON S. VEENSTRA, NARAYANAN IYER, WENJING XIE,
BENJAMIN A. LYONS, CHANG SUP PARK, AND YANG FENG

Collective action associated with social movement organizations has often been modeled as a top-down group behavior (Oberschall, 1973). Formal organizations mobilize membership and sympathetic individuals to protest by taking advantage of formal organizational ties and communicating from the organization to the public (Oliver & Marwell, 1992). This model has held true for both physical protest gatherings and other types of activist behavior organized by social movement organizations (Oliver, 1983).

This formal model is challenged by the low-cost, informal networking potential of the Internet, particularly social media platforms. These technologies allow informal groups to form and mobilize apart from formal organizations, with no continued movement structure. They also afford greater incidental participation, such as information redistribution, lobbying of public and private interests, and network bridging.

The possibilities afforded by these technologies have been most dramatically put on display in recent uprisings in several Middle Eastern countries where formal social movement organizations play a restricted role. While there is no consensus about social media's impact on these events, it is clear that the speed and distributed informality of social media allowed for quick, secret mobilization. As such, examining communication patterns within these new, sometimes ephemeral networks is a key part of reassessing our understanding of how protest movements form and sustain themselves.

This study examines the role of Twitter in the first 3 weeks of the 2011 Wisconsin labor protests, between the February 14 introduction of Governor Scott Walker's "budget repair bill" and its passage on March 11, looking at how retweeting may have been undertaken for protest mobilization, information provision, and in-group solidarity. Among other things, the bill stripped public employee unions of their collective bargaining rights. Protests against the bill topped 100,000 people and coincided with protests against similar bills in neighboring states.

During this period, Twitter users commonly used hashtags related to the bill and/or the protests, including several supporting one side in particular (e.g., **#standwithwalker** and **#killthebill**; see Veenstra, Iyer, Hossain, & Park, 2014). This study examines tweets with the more neutral **#wiunion** hashtag, which was used by both supporters and opponents of the bill. The hashtag was introduced on February 11 by Madison-based alternative weekly *Isthmus* as a way of categorizing tweets about the bill and its attendant controversy (Knutsen, 2011). The choice to restrict our examination was made under the assumption that individuals looking to Twitter for news or to get involved might use this well-publicized hashtag as a starting point. Retweeting within this network does not necessarily indicate endorsement of another's original tweet but does provide a concrete connection between the two that can be seen as part of a new, informal protest information network. That network provides the thrust of the social media challenge to the traditional model of movement mobilization.

SOCIAL MOVEMENTS AND SOCIAL MEDIA

On their face, the Wisconsin protests have much in common with previous protests at the state capitol building in Madison and sites around the world. Traditionally, such events have been organized through organizational resource mobilization, based on relatively slow communication channels, and have been largely dependent on proximity and affect (Snow, Zurcher, & Ekland-Olson, 1980).

Fundamentally, digital communication facilitates much easier *informal* communication between a movement and unaffiliated but sympathetic individuals, as well as decentralized communication between and among those individuals (Diani, 2000). Thus, the Internet provides not only a bridge from a movement to the public but also a method for engagement and participation by individuals who may never have significant contact with any formal organization, empowering an expanded network of weak ties.

In addition, for people otherwise dependent on media sources providing limited viewpoints, online news information can be "deployed in a democratic and emancipatory manner by a growing planetary citizenry," as in the early 2003 protests against the then-looming Iraq War (Kahn & Kellner, 2004, p. 88). The Internet's most important function may have been to counteract the suppressive effect of traditional media, especially television, on the likelihood of protesting (Hwang, Schmierbach, Paek, Gil de Zúñiga, & Shah, 2006).

While the use of social media to mobilize might be seen as modernized organizational communication, the inextricable link to news media and information also allows this phenomenon to be seen as a kind of citizen journalism, in which participation is a key factor in one's ability to gather and distribute information.

Since Twitter creates a mobile, real-time information network, protesters have frequently used it to remain connected and provide information to the world during protest activity (e.g., Parmalee & Bichard, 2012). Due to censorship enforced by some governments, Twitter may be a primary source of information from protests (Barnett, 2011) and provides tools for movements to circumvent the state's chokehold on information (Howard, 2010).

In 2009, Moldovans and Iranians used the service to disseminate information and arouse mass media and others' attention to protests over alleged voter fraud (Parmalee & Bichard, 2012). In 2011, the protesters in Egypt who pushed President Hosni Mubarak to step down were also aided by Twitter in disseminating mobilizing information (Cross, 2011; Parmalee & Bichard, 2012).

Beyond communication within the movement, Twitter users called for more coverage of the 2009 Iran protests by using #cnnfail to criticize the news network's reporting (Howard, 2010; Snow, 2010). Shortly afterward, CNN's Iranian coverage increased significantly (Parmalee & Bichard, 2012). This direct, informal approach to critique was made possible by the new channels Twitter has made available to the general public.

Several studies have investigated the retweet feature within large samples of tweets. In emergency situations, Starbird and Palen (2010) found that people generally retweeted information that was more topical and that non-retweets had less immediate relevance (p. 7). The same study found that messages originating from close proximity to an event were more likely to get retweeted than those from farther away (p. 7). In addition, Suh, Hong, Pirolli, and Chi (2010) found a strong link between the number of followers a user had and their likelihood of being retweeted. Likewise, messages containing hashtags and URLs were more likely to be retweeted. Finally, Romero, Meeder, and Kleinberg (2011) found messages with political hashtags were more likely to be retweeted than messages with other types of tags.

These studies together suggest multiple roles for Twitter in the process of crisis and protest communication, culminating in a many-to-many model that aggregates individuals' contributions to both circumvent and interact with traditional news sources.

RESEARCH QUESTIONS AND HYPOTHESES

Starbird and Palen (2010, 2012) suggest that both informative content and content from close to the scene of an event tend to prompt greater retweeting. Given the nature of live tweeting at events, we see nontraditional media or firsthand Twitter sources, as well as tweets from mobile phones, as indicators of close to the scene information, which might act as a proxy for the location data that not all tweets have. Thus, we propose the following hypotheses:

> *H1. Tweets that a) include news content, b) include content from nontraditional sources, and c) come from mobile phones are retweeted more frequently than others.*

Furthermore, Suh and colleagues (2010) found that the number of followers one has predicts retweeting, while Kwak, Lee, Park, and Moon (2010) suggest that this is not always the case and that it is a proxy for perceived legitimacy. Additionally, Starbird and Palen (2010) find that news organization tweets are more likely than others to be retweeted. Thus, we propose that:

> *H2. Tweets posted by organizations and their representatives are retweeted more frequently than those posted by individuals.*

Finally, following on the finding of Hansen, Arvidsson, Nielsen, Colleoni, and Etter (2011) that positive tone prompts retweeting of non-news tweets, and based on the assumption that most tweets do not contain news content, we propose that:

> *H3. Tweets with a positive tone are more frequently retweeted than others.*

To help understand the partisan nature of the fight over the bill, we additionally propose exploratory research questions about tweet content to shed light on how **#wiunion** was used and who controlled the discourse within it:

> *RQ1. Does protest support or opposition impact retweet frequency?*

> *RQ2. Does mentioning a) people on the right, or b) people on the left impact retweet frequency? Do mentions of c) the right, or d) the left interact with protest support or tweet tone?*

Starbird and Palen (2010), Vieweg, Hughes, Starbird, and Palen (2010), and Qu, Huang, Zhang, and Zhang (2011) show that situation-relevant and informative tweets are more likely to be retweeted. This is true in the specific case of constructing a protest network as well (Cross, 2011; Parmalee & Bichard, 2012). Thus, we propose the following hypotheses regarding the nature of the broader **#wiunion** protest network:

H4. Tweets that include explicit requests to be retweeted are retweeted more frequently than others.

H5. Tweets containing information intended to mobilize protest participation are retweeted more frequently than others.

H6. Tweets that mention contemporaneous protests events that are not directed at the Wisconsin budget repair bill are retweeted more frequently than others.

METHODS

This study is based on an archive of 775,030 tweets posted between February 17 and March 13, 2011, with the **#wiunion** hashtag. An archival tool, Twapper-Keeper (now part of HootSuite), was used to retrieve **#wiunion** tweets during this period, and our archive includes all such tweets from the evening of February 17 to midday on March 13, excluding a 20-hour period on March 9 during which TwapperKeeper experienced technical problems and did not archive any new **#wiunion** tweets.

From the raw data provided by TwapperKeeper, we retrieved a number of variables: the text of the tweet itself (including all tags and URLs), the username of the person who made the tweet, the name of the Twitter client used to make the tweet, and the tweet's timestamp. Additionally, we created a variable for whether the tweet was a retweet.

To examine retweeting behavior, six coders coded a sample of 1,830 original tweets (that is, those that were not already retweets), as well as any associated URLs, across 34 variables. These variables examined user characteristics, message tone, types of linked content, protest-mobilizing information, and mentions of particular people or groups. Additionally, the Twitter client used for each tweet was coded as either phone-only or computer-based/mixed-platform (31.8% phone-only). All variables are dichotomous, except protest support (support, oppose, neither). Inter-coder reliability for these measures was tested using Fleiss's kappa (Fleiss, 1971), which ranged from .36 to 1.00 (see Appendix). Nineteen values were in the "substantial" or "almost perfect" ranges of reliability, 12 in the "moderate" range, and 3 in the "fair" range, according to the Landis and Koch (1977) guidelines.

Several variables were created by collapsing multiple categories. Positivity (present in 20.3% of tweets) and negativity (30.4%) variables were created by collapsing positive tone and supportive statements and negative tone and attacking statements, respectively. A news variable (27.2%) combined news links to traditional and nontraditional sources, as well as live streaming links and firsthand accounts. An opinion variable (49.7%) combined opinion links to traditional and nontraditional sources, as well as the presence of the user's own opinion. Variables measuring traditional (9.5%) and nontraditional (21.5%) sources combined news and opinion from each type of source, including first-person accounts as nontraditional.

A variable measuring mobilization information (9.7%) combined tweets including information about protests, resource requests, information about legislative events, and information about a boycott of Scott Walker supporters. Finally, two variables were constructed to measure mentions of prominent political figures and groups—Walker, the Koch brothers, Republicans, conservatives, and the Tea Party on the right (30.3%); Barack Obama, the "Wisconsin 14" (the Democratic senators who fled the state), Democrats, liberals, and progressives on the left (8.2%).

Our criterion variable is the number of retweets a tweet received. This was determined by searching the archive for both the first 60 characters of the text of the original tweet and the original poster's username to assure positive findings were actually retweets and not simply two people posting the same headline and link; therefore, this counts only retweets made on March 13, 2011, or earlier. The result was highly skewed, ranging from 0 to 133 ($M = 1.69$, $SD = 6.73$); 62.7% of tweets were not retweeted. To address this, our analysis uses the log of the retweet count (ranging from 0 to 2.12, $M = 0.14$, $SD = 0.31$).

RESULTS

H1 and *H2* were tested using a regression model in which the hypothesized content variables, news content, and nontraditional source content predicted retweets, with opinion content and traditional source content included as controls. Nontraditional source content was a significant predictor (ß = .10, $p < .01$), supporting *H1b*, while news content (ß = -.04, n.s.) was not, leaving *H1a* unsupported. A tweet coming from a mobile phone was a significant predictor of retweeting (ß = .13, $p < .001$), supporting *H1c*. An organization posting a tweet was also a significant predictor (ß = .05, $p < .05$), supporting *H2*.

H3, *RQ1*, and *RQ2* were addressed using a non-fully factorial ANCOVA model including variables measuring protest support, tone, and mentions of the right and left. Tone variables were combined into one 3-point scale (positive [1], balanced [0], negative [-1], $M = -.10$, $SD = .65$) in this model. Protest opposition significantly predicted retweeting ($F[2, 1808] = 5.89$, $p < .01$), despite the fact that opponents (8.5% of all tweets) were a significant minority compared with supporters (50.7%) and neutral tweeters (40.7%). Positive tone was also a significant predictor of retweeting ($F[2, 1808] = 7.38$, $p < .001$), supporting *H3*. Mentioning right-wing figures led to significantly less retweeting ($F[1, 1808] = 4.24, p < .05$), but mentioning left-wing figures was not a significant predictor ($F[1, 1808] = .68$, n.s.).

Additionally, support and right-wing mentions interacted: opponents who didn't mention right-wing figures were significantly more likely to be retweeted ($F[4, 1808] = 6.62, p < .001$). A significant three-way interaction between right-wing mentions, support, and tone was also found ($F[4, 1808] = 2.46, p < .05$), such that positivity in the tweets of the previous interaction (opponents not mentioning the right) enhanced the effect even further.

H4 through *H6* were tested using a regression model containing variables related to construction of an informal social movement network: posting mobilization information, asking for retweets, and connecting Wisconsin's protests to other protest events. Asking for retweets ($ß = .05, p < .05$) and posting mobilization information ($ß = .05, p < .05$) were both significant, positive retweeting predictors, supporting *H4* and *H5*. However, referring to non-Wisconsin protest events did not predict retweeting ($ß = -.03$, n.s.), leaving *H6* unsupported.

DISCUSSION

Our findings are generally in line with existing understandings of retweeting. Positive and informative tweets were retweeted more, as were those with the legitimacy of a news or activist organization. In the context of the nascent informal protest network, mobilization information pertaining to Wisconsin-specific events was frequently retweeted; attempts to conceptually connect the protests to discrete events in other states and countries, however, did not result in significant retweeting. Additionally, though **#wiunion** has become a symbol of support for labor unions, protest support did not prompt extraordinary retweeting. Discussion of right-wing political figures did tend to be retweeted less than other tweets, which was not the case for discussion of left-wing political figures.

These findings demonstrate some limitations of the informal, decentralized movement model and the prominence of social media in it. It is hard to dispute social media's effectiveness in quickly and efficiently spreading mobilization instructions, helping to fuel protests that dominated Madison for weeks. But can this network serve the same functions as a resilient, long-term movement? The lack of significant connectivity between **#wiunion** and other events could mean that the network is wide but shallow (Krüpl, 2010). After losing the initial battle with the bill's passage, does this resource-heavy network remain as constituted, or is it reduced to a core of a highly engaged few?

If sympathetic individuals use Twitter to participate in a movement, there is potential that those individuals reach an aspirational level of participation before engaging the issue in a long-term way. This allows for a network with a high number of nodes and through which many more people may become aware of an issue but in which most nodes could disappear at any time.

However, while any given social network movement is ephemeral, it is possible that a network of networks may be sustained by interconnected cores of highly engaged individuals. This model does not promote persistent engagement, as a more formal organizational model would, but does promote greater virality. Thus, the reach and potential of such a movement network could outweigh that of a traditional organization for any given action.

Although our study provides numerous findings about Twitter's role in the protests and explores the connection between retweeting behavior and political factors, it has limitations that should be addressed. First, it focused specifically on the Wisconsin case, analyzing **#wiunion** tweets only, meaning we may have ignored other important social media sources during the protest. Given recent democratic movements around the world, it would be interesting to extend this study to other contexts and investigate how retweeting has helped organize those movements.

Second, although Twitter is effective in disseminating information and empowering people to participate in social and political events, it is not the only factor that plays a role in that process. As traditional social movement theory suggests, social events are the result of resource mobilization through multiple channels, including organizational and interpersonal communication, which we didn't take into consideration. Future studies may examine the influence of other mobilization efforts, including through traditional mass media, and investigate the interplay between Twitter use and other types of communication.

Further research should also explore the kind of messages that Twitter users are more likely to retweet in terms of individual users' behavior. What evaluative mechanisms do users have for determining which messages to retweet? Do those mechanisms interact with traditional agenda-setting processes? Understanding if there is a link between retweets, extent of media coverage, and subsequent social media discussion—in essence, a reciprocal agenda-setting model—would provide valuable insight.

Finally, it would be interesting to understand whether retweeting or sharing news content causes Twitter users to be more engaged and involved with the story and become active participants in it. The increased sharing of news within social networks may provide inadvertent exposure to news content for individuals who are otherwise disconnected from the news. Research on retweeting and information sharing within social networks should shed light on whether the behavior suggests only casual observation of that information or a deeper, sustained engagement with the content.

REFERENCES

Barnett, G. A. (2011). *Encyclopedia of social networks.* Thousand Oaks, CA: Sage Publications.

Cross, M. (2011). *Bloggerati, Twitterati: How blogs and Twitter are transforming popular culture.* Santa Barbara, CA: Praeger.

Diani, M. (2000). Social movement networks virtual and real. *Information, Communication & Society, 3*(3), 386–401.

Fleiss, J. L. (1971). Measuring nominal scale agreement among many raters. *Psychological Bulletin, 76*(5), 378–382.

Hansen, L. K., Arvidsson, A., Nielsen, F. A., Colleoni, E., & Etter, M. (2011). *Good friends, bad news—Affect and virality in Twitter.* Paper presented at the 6th International Conference, FutureTech 2011, June 28–30, Loutraki, Greece.

Howard, T. (2010). *Design to thrive: Creating social networks and online communities that last.* Burlington, MA: Morgan Kaufmann Publishers.

Hwang, H., Schmierbach, M., Paek, H.-J., Gil de Zúñiga, H., & Shah, D. V. (2006). Media dissociation, Internet use, and antiwar participation: A case study of political dissent and action against the war in Iraq. *Mass Communication & Society, 9*(4), 461–483.

Kahn, R., & Kellner, D. (2004). New media and Internet activism: From the "Battle of Seattle" to blogging. *New Media & Society, 6*(1), 87–95.

Knutsen, K. (2011). A guide to social media campaigns against Scott Walker's agenda for Wisconsin public unions. *Isthmus.* Retrieved from http://www.thedailypage.com/daily/article.php?article=32233

Krüpl, B. (2010). *Looking for the sweet spot between low-threshold participation and expressiveness in social media.* Paper presented at the Web Science Conference, April 26–27, Raleigh, NC.

Kwak, H., Lee, C., Park, H., & Moon, S. (2010). *What is Twitter, a social network or a news media?* Paper presented at the WWW 2010, April 26–30, Raleigh, NC.

Landis, J. R., & Koch, G. G. (1977). The measurement of observer agreement for categorical data. *Biometrics, 33*(1), 159–174.

Oberschall, A. (1973). *Social conflict and social movements.* Englewood Cliffs, NJ: Prentice-Hall.

Oliver, P. (1983). The mobilization of paid and volunteer activists in the neighborhood movement. In L. Kriesberg (Ed.), *Research in social movements, conflicts and change* (Vol. 5, pp. 133–170). Greenwich, CT: JAI Press.

Oliver, P., & Marwell, G. (1992). Mobilizing technologies for collective action. In A. Morris & C. Mueller (Eds.), *Frontiers of social movement theory* (pp. 251–272). New Haven, CT: Yale University Press.

Parmalee, J. H., & Bichard, S. L. (2012). *Politics and the Twitter revolution: How tweets influence the relationship between political leaders and the public.* Lanham, MD: Lexington Books.

Qu, Y., Huang, C., Zhang, P., & Zhang, J. (2011). *Microblogging after a major disaster in China: A case study of the 2010 Yushu earthquake.* Paper presented at the CSCW 2011, March 19–23, Hangzhou, China.

Romero, D. M., Meeder, B., & Kleinberg, J. (2011). *Differences in the mechanics of information diffusion across topics: Idioms, political hashtags, and complex contagion on Twitter.* Paper presented at the 20th International Conference on World Wide Web, March 28–April 1, Hyderabad, India.

Snow, D. A., Zurcher, L. A., Jr., & Ekland-Olson, S. (1980). Social networks and social movements: A microstructural approach to differential recruitment. *American Sociological Review, 45*(5), 787–801.

Snow, N. (2010). What's that chirping I hear? From the CNN effect to the Twitter effect. In Y. R. Kamalipour (Ed.), *Media, power, and politics in the digital age: The 2009 presidential election uprising in Iran* (pp. 97–104). Lanham, MD: Rowman & Littlefield.

Starbird, K., & Palen, L. (2010). *Pass it on?: Retweeting in mass emergency.* Paper presented at the 7th International ISCRAM Conference, May 2–5, Seattle, WA.

Starbird, K., & Palen, L. (2012). *(How) will the revolution be retweeted? Information diffusion and the 2011 Egyptian uprising.* Paper presented at the CSCW 2012, February 11–15, Seattle, WA.

Suh, B., Hong, L., Pirolli, P., & Chi, E. H. (2010). *What to be retweeted? Large scale analysis on factors impacting retweet in Twitter network.* Paper presented at the 2010 IEEE Second International Conference on Social Computing, August 20–22, Minneapolis, MN.

Veenstra, A. S., Iyer, N., Hossain, M. D., & Park, J. (2014). Time, place, technology: Twitter as an information source in the Wisconsin labor protests. *Computers in Human Behavior, 31*(1), 65–72.

Vieweg, S., Hughes, A. L., Starbird, K., & Palen, L. (2010). *Microblogging during two natural hazard events: What Twitter may contribute to situational awareness.* Paper presented at the CHI 2010: Crisis Informatics, April 10–15, Atlanta, GA.

Appendix: Intercoder reliability scores and code frequencies

	Fleiss's kappa	Frequency
User characteristics:		
user is an individual not tweeting on behalf of news or activist organization	0.69	88.1%
user is a news organization or tweeting on behalf of one	0.6	3.4%
user is an activist organization or tweeting on behalf of one	0.63	6.0%
Tweet tone:		
attacking a person or persons	0.5	17.0%
contains a negative phrase	0.59	24.9%
supporting/affirming a person or persons	0.47	12.8%
contains a positive phrase	0.81	14.0%
support or oppose protests, or neutral	0.44	50.7% support; 8.5% oppose
expression of user's opinion	0.41	45.4%
Manifest concepts:		
news from traditional news sources (linked or cited)	0.64	8.0%
opinion from traditional news sources (linked or cited)	0.55	1.8%
news from nontraditional sources including other individuals	0.46	11.7%

	Fleiss's kappa	Frequency
Manifest concepts:		
opinion from nontraditional sources including other individuals	0.36	7.4%
live info from news/blog/stream/etc. (linked)	0.82	5.8%
first-person eyewitness account	0.84	5.0%
protest logistics and instructions (time/place/calls to mobilize)	0.67	4.9%
Wisconsin-related protest/rally events outside Madison	0.38	3.7%
requests for resources (e.g., food, etc.)	0.73	0.6%
events in legislature (votes, hearings)	0.56	1.5%
Scott Walker	0.83	19.8%
Koch brothers	1	4.6%
Barack Obama	1	1.6%
Democrats	0.79	4.4%
Republicans	0.94	5.1%
Tea Party	0.82	5.4%
the right/right-wing/conservative	0.59	0.6%
the left/left-wing/liberal/progressive	0.79	0.8%
The "Wisconsin 14" (Democratic senators as a group)	0.5	2.3%
boycott/buycott/Koch brothers' companies/Walker-supporting companies	1	0.4%
references to non–Wisconsin-related protest places	0.82	4.2%
explicit request to retweet	0.78	1.1%

CHAPTER SEVEN

Hashtagging the Invisible: Bringing Private Experiences into Public Debate

An #outcry against Sexism in Germany

ANNA ANTONAKIS-NASHIF

VARIETIES OF FEMINISM IN HASHTAG COUNTERPUBLICS

Dominant power structures are shaping the way knowledge is being (re-)produced in our societies. Important questions for critical theory are therefore: Who is speaking and able to shape discourses? From what standpoint is s/he speaking? and (How) can communication power change these structures that are affecting individual lives? In this paper, I argue that the specific communicative dispositions of hashtags have opened up new possibilities for political participation and contestation, especially to those who feel underrepresented in a traditional media public.

Manuel Castells analyzes in his groundbreaking *Communication Power,* "Why, how and by whom power relations are constructed and exercised through the management of communication processes, and how these power relationships can be altered by social actors aiming for social change by influencing the public mind" (Castells, 2009, p. 3). In this context, the question of whether "the Internet" will have an emancipatory potential for gender relations has been debated for about a decade now by feminist scholars in the fields of political science, communication, and cultural studies.[1]

This discussion gained new popularity with the fast development of Web 2.0 technologies and "social networking sites." As Anita Harris argues, it is important to give "value" to these new forms of political participation:

> I suggest that we need to take seriously young women's styles of technology-enabled social and political engagement, as they represent new directions in activism, the construction of new participatory communities, and the development of new kinds of public selves. (Harris, 2008, p. 482)

Starting from this, I think analyzing different hashtags with feminist undertones, ones that attempt to make patriarchal structures of oppression visible, can give us insights into feminist spaces, their very differentiated debates, and new set-ups of political activism.

The most important element of the hashtags that will be analyzed here is the individualized accounts of experienced sexism. I argue that publically speaking out through these hashtags about injustices that often remain invisible reflects the desire to change realities. Since the process of narration itself becomes an action, women participating in the hashtag counterpublics should be defined as activists. As Vivienne and Burgess (2012, p. 362) state: "This sharing of personal stories in public spaces in pursuit of social change is an example of 'everyday activism.'"

While traditionally, these agents may not be regarded as political activists, it is important to remember that political participation theory has constantly broadened its scope since the 1970s, from a very limited understanding of what can be understood as political participation in democratic countries (e.g., elections) to more various forms such as demonstrations, sit-ins, etc. This new approach was primarily put forward by feminist scholars constantly pushing the boundaries to make especially women's participation visible. Harris notes that:

> Young women are underrepresented in many conventional forms of political practice and often use new technologies in under-valued ways. It is widely acknowledged that they feel more alienated from, and less entitled to participate in, formal political activities than young men, but are more likely to be engaged in informal, localized politics or social-conscience-style activism. (Harris, 2008, p. 481)

The particular genesis of the hashtags varies; they can be created as a response to violent manifestations of these structures being discussed in mainstream media, or they can come out of feminist hashtag publics themselves and then get acknowledged by traditional media.

To understand these particularities, when analyzing different hashtags, it is important to highlight the intersectional dimensions within them. According to Davis, intersectionality refers to "the interaction between gender, race, and other categories of difference in individual lives, social practices, institutional arrangements, and cultural ideologies and the outcomes of these interactions in terms of power" (Davis, 2008, p. 68).

In what we could call "hashtag herstory," it was especially feminists of color who were using hashtags to make the intersections of racism and sexism visible and bridge the gap between *White* feminist and antiracist hashtag discourses. Suey Parks, who launched **#notyourasiansidekick**, points out how hashtags can be used as a tool of decolonized practice: "We are interested in being useful, our hashtags being a tool to advance conversation and disseminate tools for decolonial action… not existing as spectacles to be consumed and treated as an end in and of themselves" (Parks, 2014).

#notyourasiansidekick was founded to "critically inspect the multiplicities of Asian American lived experience and reject assimilationist practices in order to extend solidarity to non-Asian people of color" (Parks, 2014). Raising awareness about the racist stereotypes of Asian women that persist in traditional media and affect women's lives in their social and political dimensions was only one element of this hashtag public. As we will see in the following, the complex intersections of power structures defining women's lives in different ways are not reflected when accessing mass media publics.

SUBALTERN COUNTERPUBLICS: CHALLENGING THE PRIVATE/PUBLIC DICHOTOMY?

Nancy Fraser develops her concept of "subaltern counterpublics" through a critical review of Jürgen Habermas's understanding of the bourgeois public sphere (Fraser, 1992). Her theoretical approach can be helpful to tackle the research question above, as women are—to differential degrees—still underprivileged on a political, social, and economic scale. Yet concrete effects of this structural discrimination often remain invisible to a broader public. For Fraser, subaltern counterpublics have a dual character defined by two functions: 1) spaces of regroupment and withdrawal; and 2) bases and training grounds for agitational activities directed towards wider publics (Fraser, 1992, p. 124; also quoted by Skalli, 2006, p. 37). The hashtag publics that we are looking at can be defined as a part of subaltern counterpublics: the women who launched it spoke from a position that they felt wasn't heard and communicated enough by mass media's established public sphere. They produced alternative knowledge by regrouping individual accounts under the theme of experienced sexism. In doing so, they were seizing definitional power, specifying for themselves what should be considered as "sexism."

The structurally open form of hashtags (on Twitter) allows interested people to follow them by content, not necessarily by only the people they may know. At the same time, when looking for further action, it is possible to get in contact with those who are using them. This possibility of creating new activist networks is crucial and part of a "regrouping" process.

This is a perfect example of Web 2.0 logics, where the spectator becomes him/herself the participant and author. This blurring of the lines between audience and producers (of knowledge) affects also the relations between private and public.

In feminist theory, the societal bias invested in the public/private binary defines spaces in society with specific gender roles. As Loubna Skalli puts it, especially for the context of MENA[2] countries, where gendered spaces are still particularly affecting political and social life, "My use of the public sphere in the context of the MENA region refers to open discursive spaces that include 'subaltern Counter publics,' where subordinated groups such as women challenge the patriarchal public/private division" (2006, p. 37).

While the feminist slogan shouted on German streets in the 70s was "The private is public!," new social media gave this slogan a novel turn, as women got access to technologies that allow them to bring their narration into public without leaving their secured space. They can wait, follow, and see the development of a hashtag, get inspired, and finally decide to take part in the discourse. Risks of being exposed when speaking up about sensitive issues are drastically reduced.

We will see by entering the "wider publics" (what I call the "spillover" from hashtag counterpublics to broader publics created by "traditional media"), that the German hashtag #aufschrei lost some of its subversive potential due to different strategies—namely, ridiculing, derailment, and direct attacks/fear—that have been used to silence the outcry both on- and offline.

#OUTCRY ON TWITTER—SEIZING NORMATIVE POWER

Outcry was triggered by a blog post where a number of young feminist activists shared their experiences. They later decided to move the conversation to the shorter form of the hashtag. Then, on the night of January 25–26, 2013, feminist and activist Anne Wizorek suggested collecting these tweets under the tag #aufschrei ("outcry").[3]

In retrospect, she places great importance on the time dimension of tweeting out the hashtag. At night, the German "feminist Twitter community bubble" was awake and kept the hashtag going. She was afraid that the next day, the hashtag would not make it on its own, that it would quickly lose attention or would not make it out of this very specific counterpublic on Twitter. Yet to her surprise, within hours, hundreds of women joined in by tweeting their experiences with everyday sexism, sexual harassment, and sexist behavior in the workplace, on public transport, or within the family. Their spontaneous accounts bring individual private experiences into a broader counterpublic.

As for the content of the 140-character outcries, they were covering a broad spectrum of testimonies about, for example, physical violence, rape, and verbal abuse. The perpetrators were often in positions of power: older male family members, teachers, bosses, etc. It is this diversity of accounts, the mass of situations described—ranging from experiences in kindergarten to those in workplaces—that made the hashtag a unique space of discussion on sexism.

Anne stressed that there was no community outreach or call to share their experiences, but women (mostly) felt moved and empowered to do so. This brings us back to the hashtag design that is taking over the function of a space of regroupment, or as Sophia Seawell puts it: "Twitter allows, embraces and encourages a plurality of messages. Hashtags, on the other hand, allow messages to be both individual and collective; there is no need for a 'unifying' message because the hashtag is designed to provide it" (Seawell, 2014). Anne's account suggests that the collective experience was also the most important outcome of the growing flood of tweets that created a collective identity within these very first hours of the night: "This moment of collective awakening: shit, that's our reality and it happens to all of us! It was freeing but hurtful at the same time. It feels good to know that you are not alone with these experiences."

This empowering feeling was created by the many who seized the power to define what has been perceived by them as sexist behavior, creating alternative knowledge. The **#outcry** broke a taboo that was being adhered to by typical statements women have internalized in their real lives, such as "don't exaggerate/this is only in your head," "you are a prude/a killjoy," etc. The masses of tweets helped many women (including those who were never in touch with feminist theory or thoughts before) to leave behind these internalized mechanisms of silencing their testimonies.

At the same time, there were also voices that pushed reflections further and added other axes of discrimination to **#outcry**, saying it mainly reflects the realities of *White* women. Kübra Gümüsay, a journalist and author of Turkish origins, launched **#schauhin** ("look closely") in order to make racist structures visible in German society. She says that the hashtag could benefit from the huge mediated success of **#outcry**, because the initiators of the latter also supported and shared **#schauhin**. While following what happened in the **#outcry** counterpublic, Kübra explains in a blogpost that she just couldn't feel the same passion, the moment of a "collective awakening," in reading the accounts. "I couldn't really understand why. Maybe because I experience sexism always in combination with racism" (Gümüsay, 2013). She was not alone in her intersectional critique—lesbians and trans persons pointed to a heteronormative bias of **#outcry** and its reception and started **#queeraufschrei** as a reaction to it. The hashtag **#schauhin** is still active; while never having experienced the same "hype" as **#outcry**, it has opened up new spaces for debate on racism in Germany.

HIJACKING HASHTAGS

The hashtag is still being used today and has resisted attempts to hijack and troll it by those who are trying to regain control over definitional power. There are multiple strategies to hijack a hashtag that was created in order to make power structures in a society visible. In her study on gender-based movements, Myra Marx Ferree (2012) has identified ridiculing,[4] victim blaming, and victimizing as three strategies of "soft repression" to silence women's protest movements in general. It's interesting to apply this typology to the specific communication context of hashtag publics. The hijack of a hashtag also aims at *silencing* those women who participate and exchange ideas via their hashtag by taking over their public.[5] I will give empirical examples for these strategies being reproduced for **#outcry's** hashtag public.

Soon after the launching of the hashtag, sexist jokes were quickly posted in this counterpublic to ridicule the movement, and got a non-negligible number of retweets. Over time, the hashtag was misused for any ironic situation in daily life to show indignation, such as "Coffee hot! **#outcry!**". The absurdity was intended to make people laugh and ridicule the intention of the hashtag. **#Outcry** became also associated with hysteria, reproducing the old stereotype of the emotionally driven fury. Tweets for victim blaming were also found on the hashtag: for example, a user identifiable as male by his profile wrote, "Mimimimi, I'm getting harassed, when wearing my transparent shirt. **#outcry**," adding the hashtag **#schwachsinssdebatte** ("stupid debate") to it.

Moreover, in the mainstream media, victimizing strategies were used on talk shows and in counterpublics alike. Tweets such as "How am I then supposed to flirt," are shifting the focus from the perspective of women to men, who seem unable to cope with the current discourse. This victimizing strategy helps them avoid reflecting on their own privilege or their own behavior towards women. Another young male user complains in a tweet: "In my office there are 34 women and 5 men. Where is equality here? **#aufschrei**." The writer (willfully) neglects the power dimension of patriarchal structures when it comes to the professional world and labor.

The anonymity of the net also offers new options for direct attacks. Anne, who has revealed her real name and face in the mainstream media, has had to face criticism and defamation, and women who shared their stories also have been the target of insults. Antifeminist and misogynist comments were directed towards them, and often they were attacked on a very personal level. This verbal abuse online is yet to be understood as a form of violence. For initiators of feminist hashtags, it is not "virtual"; it affects them as "real persons." Anne says:

Internet is my *Lebensraum*, my home. We shouldn't treat this question as if there are some idiots in the net and we should get over it. That's not true, they are symptoms of the status quo in our society. I felt: They want to get me out of there, but if I disappear, they won. But of course, it does something to you.

Against the background of the previously mentioned attempts to hijack the hashtag public, the initiators of **#outcry** took action: they created a website where experiences could be posted anonymously. In order to spread awareness of the existence of the website, on various occasions they posted the domain with the hashtag itself.

The decision to bring the hashtag into a website (alltagssexismus.de) can be read as the wish to put the debate on a more institutionalized level and to create a "safer" space (comments are moderated). At the same time, this platform was explicitly inviting people to write about their experiences with racism, homo- and transphobia, and other structures of oppression, to establish a more intersectional approach, assuming that individual experiences are being framed *not only* by patriarchal power relations *but also* by their intersections with other oppressive power structures.

These attempts to hijack hashtags are more recently being encountered by feminist tech projects such as "Zero Trollerance" (www.zerotrollerance.guru), a project of the Peng!Collective, a "subversive campaigning group." In response to the increasing levels of gender-based trolling and abuse on Twitter and the platform's lack of action against it, a task force of Twitterbots was programmed for one week to detect sexist and transphobic language on the platform and respond by enrolling the Twitter users in a self-help program designed for trolls. The "program" includes six steps and runs for 6 days. Trolls receive a playful video and tutorial each day, as well as tweets from the "coaches" (bots).

While those initiatives are designed to create safer spaces (for women) online, the immediate popularity of hashtags and the spillover from counter- to broader publics are often increasing the visibility of the women who initiated the hashtags. This abrupt publicness is not easy to handle for participants and initiators alike. At the same time, aren't these broader publics needed in order to create social change?

SPILLOVER EFFECTS—FOR THE BETTER?

The German hashtag **#outcry** trended and also reached a broader public, which means that traditional media "discovered" the problematic of sexism in the German society—"Germany discusses sexism" had become the chorus of many headlines by the beginning of 2013. But the spillover wasn't simply a recapitulation of the hashtag into traditional media.[6] The popularity of **#outcry** reached by a spillover from the hashtag public to the traditional media took a specific turn. While the initiators of **#outcry** didn't intend it, the debate got linked to an isolated case: the matter of the politician Rainer Brüderle, who was accused by a female journalist of having made sexist statements towards her. Her journalistic piece, which appeared in the high-circulation *Stern* magazine, coincided with the genesis of the hashtag. For the initiators of **#outcry**, it is clear that if the hashtag hadn't existed, this affair would have stayed in a political, isolated context. It was the combination of the **#outcry** trend in the hashtag public on the one hand and the famous politician associated with sexist behavior on the other that provoked a nationwide debate. At the same time, when speaking on talk shows with other participants, it was difficult for the initiators of **#outcry** to broaden the frame of the debate by overcoming the focus on the Brüderle affair and discussing the structural dimensions of sexism.

In June 2013, **#aufschrei** won the renowned online media award Grimme. For the first time, the prize was dedicated to a hashtag: the jury recognized the important effects that the hashtag had on the dynamics of a societal debate. At the same time, by winning the award, the hashtag seemingly was trapped in the logic of a debate with a start and an endpoint. On the contrary: due to the complexity of power relations that can be challenged by feminist counterpublics, the activist hashtags are being created to endure.

With the example of **#outcry** to hand, we can also see how in Germany, the experiences with sexism have remained more of a *White* debate, as if sexual harassment is something that happens only to *White* women. We have seen how women affected by racism were little represented by tweeting their experiences with the interlocking systems of patriarchy and racism (and/or heteronormativity) on this hashtag. Anne, a *White* feminist, was conscious of these issues, but "When I tried to make clear to journalists that I'm not talking only about a *White* phenomenon, and I tried to make clear the complexity and intersections that affect women's lives, these were usually the sentences that were kicked out first."

When entering mainstream media, the logics become different, the discourse gets abstracted and less subversive. A nonhierarchical outcry pronounced by the many without a spokesperson became centered around only one woman[7]—Anne, who then had to speak for all women in mainstream media and to educate people about new social media at the same time. "It was hard for me to talk about feminisms and social media with such broad publics," says Anne, continuing, "I had to explain to some journalists what a hashtag is; a lot thought that it's *me* who is controlling the hashtag." This is understandable, as we have seen, due to the generation gap (the majority of TV spectators of one of the most famous talk shows on ARD, where Anne was invited, are over 40 years old). With her recently published book *Ein Aufschrei ist nicht genug* (*One Outcry Is Not Enough*, 2014), she also engages other publics through print media.

Besides this technical knowledge determinant, the understanding of what is happening inside a counterpublic was limited. As Bianca Fritz concludes in her analysis on the reception of **#outcry** in the mass media, "The German society is not trained in integrating the generated knowledge (of outcry on Twitter)" (Fritz, 2014, p. 1). The initiators of **#outcry** point to the trivialization of the problem of sexism as soon as it becomes a topic for a TV talk show, where they are discussing questions such as "Is it then still okay to open the door for a woman, or will this be perceived as sexism?" in order not to look at the structural dimensions of the experiences of sexism that the hashtag is intended to denounce. One argument in German discourse references the U.S., "where men can get sued just because they use the same elevator as a woman." This can also be seen as a strategy of derailment, as the analytical focus is shifted from their own society to another.

Young feminists in Germany feel particularly voiceless: the mere arrival of Chancellor Angela Merkel seems to have justified the claim that Germany is the example of a perfectly equal society. As a consequence, the discourse around **#aufschrei** in the broader public has stayed on a rather superficial, entertaining level.

As Suey Park states: "The media makes our movements into spectacles, rather than acknowledging them as the origins of serious decrees for radical action" (Parks, 2014).

Rather than being useful in getting their cause into a broader public, the anticolonial project **#notyourasiansidekick** is being "co-opted" and is losing its subversive potential as it becomes simply entertaining. The traditional mechanisms of societal power relations that dominate the one-to-many media landscape are then being applied once again.

Dealing with different publics seems to require different strategies from the activists. Interestingly, the 1-year anniversary of the **#outcry** debate has been celebrated with a flood of articles, resuming the campaign and its effects and prompting interviews with the founder that mainly ask whether "it has changed something in the lives of women." The government-funded magazine *Aus Politik und Zeitgeschichte*[8] dedicated a whole edition to the topic of sexism following **#outcry**, announcing that the authors would take a more "scientific" approach to the debate that had swept across the nation 1 year before. One text (Bönt, 2014) illustrates well how to downplay the hashtag public that has been created: first of all, the author denies that there was a "debate," and understands the thousands of tweets as "entertainment." Terms like "scream queen" that the author uses in his text to describe the hashtag's participants imply the stereotype of the hysterical woman. In the text, we also find the *Ablenkungsstrategie* ("distraction strategy") derailment, which turns the focus from women and their experiences with everyday sexism to the (surely also important) problems of family fathers. On another level, he also uses typical strategies of victimization in his article. From this we can conclude here that even when the hashtag public entered the academic discourse more than a year later, it wasn't immune from trolling attempts.

SIMILAR OBJECTIVES AND CHALLENGES: "NOW CAN WE GET SOME *REAL* SOCIAL CHANGE?"[9]

We can state that hashtag publics are bound by language barriers. Having analyzed in depth the particular "national" hashtag **#aufschrei**, where the language of communication was German,[10] you can of course see comparable hashtag publics (temporarily) emerging in other countries or language-defined spaces. In drawing these parallels, I don't by any means mean to put all of these hashtags and the specific struggles they represent into one box. There are many factors influencing the trending possibilities of feminist hashtags that should be taken into account: political and socioeconomic contexts, Internet penetration and literacy rates, and media and censorship systems determine the possibilities for social change of hashtag publics. At the same time, it is interesting to note that small feminist hashtag publics are popping up all around the world to challenge patriarchal systems using analogue methods.

The hashtag **#weareallwomen** is an example from 2014 with a similar objective: bringing in the perspective of women. The genesis of this hashtag was much more violent; it can be read as a response to the misogynist assassination of seven women in the U.S. by a young man who had publically stated his hatred of women. What followed were tens of thousands of accounts by women illustrating the structures that had made this hate crime possible. The traditional media lacked any critical reflection on the underlying structures of this hate crime; while they portrayed it as an exceptional case of a "sick" person, the discourse in those counterpublics gave insights into the systematic dimension of this tragedy. We have seen in the German debate that isolating these incidents and reducing them to individual cases stops a societal debate on the structural origins of crimes committed against women. This example shows the reciprocity of spillover effects: they flow not only from hashtag publics into traditional media publics but also vice versa. In this particular context, the counterpublic also had a corrective role in differentiating the mainstream discourse.

In Tunisia, the hashtag **#moi_aussi_j'ai_**été_violenté ("me too, I have been raped") was launched by a small group of feminist activists in the capital. Here too, a murderous crime against a woman triggered the hashtag, which was present on Facebook, not on Twitter.[11] The case of a father who burned alive his 14-year-old daughter because she had brought a male college student home sparked a debate in the media and civil society. The number of women who decided to share their experiences under this hashtag in order to regroup and give the topic a more structural societal dimension was very limited, though. This was due to the fact that there was no spillover effect into traditional media, as we have seen in the German case. The initiator, Amal, was invited for a radio interview, but it was clear that decision makers in the national media weren't willing to give this subject more space. She brings attention to the fact that keeping a hashtag going often demands various resources from the activists. In another example, the hashtag **#Taharush** ("harassment") was launched in Egypt and was and is used extensively during protests. It is impossible to enumerate here all of the various examples of feminist hashtags all over the globe. For scholars, it is hard to cope with all of these counterpublics, as many only come to the light when they reach a "critical mass" of tweets or after having spilled over to traditional media. In addition, a regional expertise is needed to contextualize the hashtags and their potential for social change.

The classic question to investigate these new activist spaces would be, can these specific hashtag publics bring about social change? The "outcomes" certainly vary drastically. While feminist activist and journalist Henda Hendoud states that in the Tunisian context, they represent a necessary step to break taboos in society, Anne points to a concrete effect of **#outcry** in the German context: the state's anti-discrimination facilities saw a rise in demand by one third after **#outcry** swept onto the national scene, and as a consequence, services have been diversified. Already existing organizations and NGOs have noticed that they have to increase their visibility and to offer more antiharassment training, for example, in companies and state institutions.

At the same time, the number of people "using social media as activist politic" (Parks, 2014) is growing. The dichotomies of virtual/real, public/private, and talk/action may still serve as analytical tools, but they have little to do with empirical evidence. Against this background, more research needs to be done in the field of reframing social change, "outcomes," and "effects" of political participation of different actors on social media (see, for example, Tufecki, 2014). While feminist hashtag publics constitute empowering spaces of exchange where dominant power relations and their intersections can be made visible and criticized, it seems that a reliable translation of such discourses to a broader public is not yet possible.

NOTES

1. For the German academic debate on gendered political communication, see, e.g., Harders & Hesse, 2006.
2. MENA stands for Middle East and North African countries. While this distinction has been criticized for constructing a geographical space based on racist assumptions that detach the

northern part of Africa from the rest of the continent, I still employ it here as a language-defined space.

3. **#aufschrei** will be translated in the following as **#outcry**. The public under investigation here was constructed around the German word only, though.

4. In Myra Marx Ferree's study on gender-based movements, "ridicule" has been identified as a category of "soft repression" to silence women's protest movements. It can be interesting to apply this category in the specific communication context of hashtag publics.

5. For another example of attempts to hijack or swamp hashtags to silence critiques of sexism by women, see the discussion of **#GamerGate** in Rambukkana, Chapter 2, this volume.

6. I argue that a media system that usually outsources the issue at hand to other countries, such as the "Arab World," couldn't resist the thousands of tweets.

7. The two other initiators, one of them writing for the famous feminist blog *Mädchenmannschaft*, did also give interviews but lacked the TV presence.

8. You can find the dossier online: http://www.bpb.de/apuz/178658/sexismus. All translations have been made by the author.

9. Kai Ma asks this question in her piece on Time.com, in a critical reflection of the hashtag **#notyourasiansidekick**. http://ideas.time.com/2013/12/18/notyourasiansidekick-is-great-now-can-we-get-some-real-social-change/

10. Anne states in the interview that it was important to her to choose a German name for the hashtag, as feminist counterpublic discourses in Germany are very much influenced by English language, excluding women who don't speak English (well) and don't feel comfortable with these terminologies. Until now, key concepts such as "victim blaming" are still lacking an adequate translation.

11. Facebook is much more used than Twitter in Tunisia. In my interviews, the Twitter community is referred to as an "exclusive elitist circle." For statistics, see the annual Arab Social Media Report, http://www.arabsocialmediareport.com/home/index.aspx

REFERENCES

Baym, N., & boyd, d. (2012). Socially mediated publicness: An introduction. *Journal of Broadcasting & Electronic Media, 56*(3), 320–329. doi:http://dx.doi.org/10.1080/08838151.2012.705200

Bönt, R. (2014). Tausendschön im Neopatriarchat (B. f. Bildung, Ed.) *Aus Politik und Zeitgeschichte.* Retrieved from http://www.bpb.de/apuz/178666/tausendschoen-im-neopatriarchat

Castells, M. (2009). *Communication power.* Oxford, UK: Oxford University Press.

Consalvo, M., & Paasonen, S. (Eds.). (2002). *Women and everyday uses of the Internet: Agency and identity* (8th ed., Vol. Digital Formations). New York: Peter Lang.

Davis, K. (2008). Intersectionality as buzzword: A sociology of science perspective on what makes a feminist theory successful. *Feminist Theory, 9*(1), 67–85.

Dubai School of Government. (2011, May). Civil movements: The impact of Facebook and Twitter. *Arab Social Media Report, 1*(2). Retrieved from http://unpan1.un.org/intradoc/groups/public/documents/dsg/unpan050860.pdf

Eimeren, B. van, & Frees, B. (2013). Rasanter Anstieg des Internetkonsums – Onliner fast drei Stunden täglich im Netz. *Media Perspektiven, 7–8*, 358–372. Retrieved from http://www.ard-zdf-onlinestudie.de/index.php?id=415

Fraser, N. (1992). Rethinking the public sphere: A contribution to the critique of actually existing democracy. In C. Calhoun (Ed.), *Habermas and the public sphere* (pp. 109–142). London: MIT Press.

Fritz, B. (2014). Ein Aufschrei im Internet, der im TV verhallt? Diskursanalytische Annäherung an die Sexismus-Debatte in Deutschland 2013. Master thesis. University of Basel, Switzerland. Retrieved from http://www.biancafritz.com/wissenschaftliches/ abgerufen

Gümüsay, K. (2013, September 9). #SchauHin – der neue #Aufschrei. *MigaZin: Migration in Germany*. Retrieved from http://www.migazin.de/2013/09/09/schauhin-der-neue-aufschrei

Harders, C., & Hesse, F. (2006). Geschlechterverhältnisse in der Blogosphäre: Die Bedeutung der Kategorie Geschlecht für die Verwirklichung von Teilhabechancen durch neue Medien. *Femina Politica* (2), 90–101.

Harris, A. (2008). Young women, late modern politics, and the participatory possibilities of online cultures. *Journal of Youth Studies, 11*(5), 481–495.

Mai, K. (2013, December 18): #NotYourAsianSidekick is great. Now can we get some real social change? *Time*. Retrieved from http://ideas.time.com/2013/12/18/notyourasiansidekick-is-great-now-can-we-get-some-real-social-change/

Marx Ferree, M. (2012). *Varieties of feminism*. Stanford, CA: Stanford University Press.

Parks, S. (2014). Hashtags as decolonial projects with radical origins. *Model view culture: Technology, culture and diversity media*. Retrieved from https://modelviewculture.com/pieces/hashtags-as-decolonial-projects-with-radical-origins

Seawell, S. (2014). #NotYourAsianSidekick: Rethinking protest spaces and tactics. *HASTAC—Humanities, Arts, Science, and Technology Alliance and Collaboratory*. Retrieved from http://www.hastac.org/blogs/sseawell/2014/03/12/notyourasiansidekick-rethinking-protest-spaces-and-tactics

Skalli, L. H. (2006). Communicating gender in the public sphere: Women and information technologies in the MENA. *Journal of Middle East Women's Studies, 2*, 35–59.

Tufecki, Z. (2014). Social movements and government's response in the digital age: Evaluating a complex landscape. *Journal of International Affairs, 68*(1), 1–18.

Vivienne, S., & Burgess, J. (2012). The digital storyteller's stage: Queer everyday activists negotiating privacy and publicness. *Journal of Broadcasting & Electronic Media, 3*, 362–377.

Wizorek, A. (2014). *Weil ein Aufschrei nicht reicht: Für einen Feminismus von heute*. Waldachtal-Tumlingen, Germany: Fischer.

CHAPTER EIGHT

Hashtags as Intermedia Agency Resources before FIFA World Cup 2014 in Brazil[1]

CARLOS D'ANDRÉA, GEANE ALZAMORA, AND JOANA ZILLER

INTRODUCTION

In June 2013, protests erupted in several Brazilian cities during the Confederations Cup as a part of an intense political demonstration. Triggered by an increase in bus fares in São Paulo, these protests quickly highlighted new issues, such as the overexpenditure associated with the 2014 FIFA World Cup and governmental corruption.

Similar to people in other countries like Spain (during the 15M movement in 2011), and Egypt, Tunisia, and others (during Arab Spring in 2010–2012), the Brazilian people planned and communicated about these street protests primarily using online social networks such as Twitter, Facebook, and Instagram, evincing a complex interconnection between urban space and cyberspace. The most popularly utilized hashtag, **#vemprarua** ("come to the street") demonstrates this. This hashtag created a hybridity between urban space and the Internet using online social networks to nudge the Brazilians to abandon their "slacktivism."

After the Confederations Cup, the intense street protests transformed into a continuous discussion about the effects and consequences of holding the FIFA World Cup in the country. Social groups aiming to condemn the atrocities committed in the name of such a mega sporting event organized themselves around the phrase "there will be no World Cup," which quickly branded the tweets, posts, and urban interventions with the hashtag **#naovaitercopa**.

This slogan/hashtag also referred to the street demonstrations that often culminated in clashes between the demonstrators and the police. Concerned with the negative impact of face-to-face and virtual events, the federal government attempted to express a positive agenda through slogans such as "World Cup of the World Cups" (#copadascopas) and "there will be a World Cup" (#vaitercopa), resulting in a process we refer to as "hashtag war."

Apart from being the link between street protests, online social networks, and political discussions, the hashtags were also appropriated for other additional purposes. For example, the hashtag #naovaitercopa resulted in several ironic memes that referred to current events that were less related to political issues. Furthermore, advertising campaigns organized by World Cup sponsor companies sought out slogan/hashtags such as #issomudaojogo ("it changes the game") to associate their brands with both the World Cup and the political "changes" that were underway in the country. Marked by record numbers in the streets and their sharing or production for online circulation, the June Journeys in Brazil and the subsequent events can be considered a typical example of a convergence culture (Jenkins, 2006), creating an amalgam of elements such as commercial mass media, social media, fashion, and street art.

The omnipresence of a graphic character # could be noticed both online and on the streets. Having initially appeared in posts and tweets on social networks, the hashtags began to act as a synthesis of various collective demands and came to denote a "war cry" heard throughout cities and digital networks. In urban spaces, on the Internet, and even in mass media, the hashtags are seen here as one of the promoters of the sociopolitical controversies that have gained prominence in light of this mega sporting event in Brazil.

In this context, our study aims to explain and analyze the characteristics and specificities of the hashtags that were created during the 2013 June Journeys, as well as to investigate the political and communicational events that took place after this mass event in Brazil. We believe that street events, online mobilizations, advertising campaigns, and other appropriations against and in favor of the 2014 FIFA World Cup propagated through networks articulated by elements such as hashtags, which can be viewed as intermedia agency resources (Callon, 2006; Herkman, 2012).

To achieve the aims of this study, we employed the perspectives provided by actor-network theory (Callon, 2006; Latour, 2005) based in Deleuze and Guattari's philosophies. The ideas of Bruno Latour and Jacques Rancière are also used here to discuss politics as assemblages marked by heterogeneity and for ephemerality. We also discuss how intermedia connections around hashtags expand their social reach and promote remixes of the typical meme's logic (Bauckhage, 2011; Jenkins, 2013).

#VEMPRARUA: BETWEEN STREETS AND SCREENS

In June 2013, on the eve of the FIFA Confederations Cup, hundreds of Brazilian cities experienced intensive political demonstrations. In spite of the previous climate of dissatisfaction with the expenditure and the restrictions imposed by the 2014 FIFA World Cup, it is evident that the "ground zero" of June Journeys was the demonstrations led by *movimento passe livre* (free fare movement) against the U\$0.10 price increase of bus tickets in São Paulo. Strong police repression against the organized June 13 street rally triggered other protests in many state capitals, and a few days later it spread to hundreds of Brazilian cities.

Until the end of the preparatory sports competition for the 2014 FIFA World Cup on June 30, hundreds of thousands of people took to the streets every day, leveraging the country's largest political event in at least 20 years. Besides the opposition to overspending for the mega sporting event and government corruption, there existed criticism of the press (primarily TV stations) and political parties, as well as a refusal of traditional social movements.[2]

The adoption of hashtags was organized in a bottom-up fashion, and some variations were used in a self-organized way at each new protest in different cities. With no central command, the protests were scheduled through Facebook events; frequently, more than one event was created for each protest by different people. However, it was through hashtags that ad hoc communities (Bruns & Burgess, 2011, Chapter 1, this volume) began to form. These agency resources allowed people to discuss common themes, share feelings, and call for street protests, as well as publicize the reports recorded from them.

The #vemprarua ("come to the street") hashtag, one of the most commonly used by demonstrators who protested during the Confederations Cup, is illustrative in the dynamics of entanglement between streets and social media that characterized the June Journeys. It is herein assumed that the Brazilian protests of June 2013 worked as networked events that influenced both social media and the streets, without being simplistically reduced to either one. Such events are governed by the logic of connections (Kastrup, 2004), which integrate both online and offline devices in a social-communicational dynamic that is deeply affected by contemporary media processes.

This interlacing is made clear, for example, when we identify that users of social networks also utilize hashtags offline, during everyday life. During demonstrations, handwritten posters (see Figure 4.1) and shirts carrying slogans preceded by hashtagging could be seen, and after the conclusion of the street protests, it was easy to identify graffiti and other forms of street art marked with hashtags that frequently referred to trending topics.[3] While "walking" through the city, hashtags originating from Twitter, Facebook, and Instagram were renewed in their circulation context. While a hashtag mobilized through an online network appears to be

more temporally specific, its inscription throughout the city has an enduring physical character and reaches a different audience.[4] Thus, a hashtag such as **#vempra-
rua** is posted, commented on, and shared not only through social networking, but
also throughout urban and intermedia connections, which considerably broadens
its agency.

Figure 4.1. Protesters with shirts displaying the **#vemprarua** hashtag during a demonstration in
Belo Horizonte (image by Carlos d'Andrea, June 2013).

INTERMEDIA AGENCY AND POLITICS

Social media, or society, as explained by Latour (2005), is not something that
exists a priori and as such cannot be understood in its static or finished form. As
explained by actor-network theory, its understanding depends on the realization of
how the different actors—both human and nonhuman—act constantly and influence others who make up a network that is always changing.

In this work, we understand the street protests and mobilizations underway
in Brazil since June 2013 via the Internet as the result of hybrid and collective
agencies that give visibility to fairly heterogeneous demands articulated by temporary groups. Whereas "anything that does modify the state of affairs by making a
difference is an actor" (Latour, 2005, p. 71), it seems clear that hashtags, through
their online social networks and adaptation into urban space, are one of the central
features of the articulation of June Journeys and its offshoots.

If *agencement* assumes the ability to act in various ways, depending on the agency configuration (Callon, 2006), the socio-technical *agencement* "includes the statement[s] pointing to it, and it is because the former includes the latter that the *agencement* acts in line with the statement, just as the operating instructions are part of the device and participate in making it work" (Callon, 2006, p. 13). Social and political implications of collective *agencement* concerning hashtags are therefore intermediately woven through shares.

Whereas "intermediality asserts that political communication takes place by increasing the number of media channels and communication technologies, which are inherently linked to each other" (Herkman, 2012, p. 370), we consider the intermedia connections regarding hashtags within the framework of June Journeys as a large-scale political and socio-communicational dynamic. The more recognition the hashtag **#vemprarua** received intermedia connections, for example, the more it brokered political actions among the demonstrators in the country's streets.

The logic of collective action, or networked action logic (Bennett & Segerberg, 2012), involves paradigmatic changes in social and political relations by means of dynamic intermedia sharing. "In this connective logic, taking public action or contributing to a common good becomes an act of personal expression and recognition or self-validation achieved by sharing ideas and actions in trusted relationships" (Bennett & Segerberg, 2012, p. 752).

Associated with various claims, hashtags like **#vemprarua** mobilized a certain political attitude during the June Journeys, integrating a kind of collective action into a network dynamic. According to Latour (2003), the heterogeneous mediations—contaminations or connections—have regrouped to form provisional coherences related to the schemes of enunciation, as with political enunciation. For that, Latour claims that the agents need to change their opinion, and those belongings may change throughout the debate. "It is in this slight dislocation of discourse that the mini-transcendence of politics lies, the one that enables it to agitate public life, to cause it to ferment, sometimes to boil, to disrupt but also to 'stir' it, so to say, before clarifying it, in any case to increase its temperature" (Latour, 2003, p. 160).

By agreeing with Latour (2003), we consider that the collective and diverse use of hashtags in social-technical arrangements that permeated June Journeys in 2013 in Brazil had no function to unify the political discourse on such a porous interface between streets and social media. On the contrary, hashtags have contributed to the management of interim understandings, which were often contradictory. And it is precisely these aspects, which are rarely versed in traditional political representation, that make them deeply political and attuned to the contemporary character of collective assemblages.

In the same sense, according to Rancière (2008), isolated, collective, and even contradictory actions of a circumstantial and ephemeral group characterize politics and not the common statement. This is the inconsistent character of convergence, which to us seems more appropriate for addressing the June Journeys intermedia circulation. Such heterogeneity and its associated power relations become more evident if we problematize the proliferation of local variations of the hashtag #protestos. Each city gets a formal hashtag: #protestosp for São Paulo, #protestobh for Belo Horizonte, #protestorj for Rio de Janeiro, etc.

After monitoring 35 Twitter hashtags and keywords associated with the protests, Bastos, Recuero, and Zago (2014) conducted a study based on the geographical location of tweets (they identified the geographic location of almost 50% of the dataset, or 1.4 million tweets). In their study, the authors identified that a significant volume of tweets with hashtags such as #protestorj and #protestosp did not necessarily originate from the residents of those towns but rather from inhabitants of more peripheral regions in Brazil. Overall, tweets about protests "were identified as coming from or referring to 3,268 Brazilian cities across the 27 federative units." They continue, "users' hashtag messages aimed at drawing attention to a particular cause or opinion regardless of whether they are physically present at that location" (2014). In one of the study conclusions, the authors point out that "hashtags thus connect regions geographically more isolated to urban centers while simultaneously offering a platform that brings social media users closer to street protests" (2014).

In this context, it's important to point out the diversity of hashtags used in regard to Brazilian protests in June to help us to understand that, while these people were protesting together, they were motivated by specific causes, making a clear "distribution of the sensible" (Rancière, 2008, p. 12). The "communities of sharing" engineered by hashtags were organized simultaneously due to articulations between digital networks and urban space; they were also organized based on an identification with the causes and the ongoing political actions, regardless of the locations of ad hoc members of the temporary groups.

"THERE WILL BE NO WORLD CUP"?: A HASHTAG WAR

Amidst the numerous demands and raised flags during June Journeys, a "war cry" seemed to progress and linger during the time between the Confederations Cup and the FIFA World Cup. "There will be no World Cup," which in Portuguese was "Não vai ter Copa," became one of the binding expressions of the opposing interests catalyzed by June Journeys' protests.

For some time, the effective realization of the World Cup seemed threatened by the country's momentary political instability. When large protests had thinned out, however, the expression/hashtag #naovaitercopa had gained a metaphorical meaning: although the sporting event was in fact going to happen, a significant portion of the population would not share the joy and camaraderie promised in the advertising campaigns produced by the government and FIFA.

The hashtag **#naovaitercopa** thus came to be associated with a wide range of demands and expectations of very heterogeneous social groups. One of the strongest appropriations came from social mobilizations expressed by the impact the World Cup had on the host cities. In several localities, "popular cup committees" were organized to accompany the urban and social impacts of the mega event, especially on the poorest people. Among other situations, thousands of people were removed from their homes for the expansion of the roads, construction of stadiums, and other works connected to the World Cup. The expression "there will be no World Cup" was also used to christen the street acts in São Paulo in the first half of 2014.

For this group of activists, other hashtags were relevant, such as "World Cup for whom?" (in Portuguese, **#copapraquem**). A Facebook post published in February 2014 by the Atingidos pela Copa Committee, for example, displays this hashtag in the footer. The title poses a question preceded by a #: "Won't we have housing?"[5]

The expression **#nãovaitercopa** also began to be associated with the broader critical issues, such as the problems that arose with the country's infrastructure, corruption (especially in the public sector), and the distrust in politics—particularly with the Workers' Party, which had presided over the country with Luiz Inácio Lula da Silva since 2002. The current president of the Republic, Dilma Rousseff, was one of the protestors' prime targets, and the protests culminated in a certain anticipation of discussions around the presidential elections scheduled for October 2014, for which the president was regarded as a favorite for re-election.[6] In November 2013, for example, the president's critics and opponents had given some attention to the hashtag **#BlocoQuemTemBocaVaiaDilma** (which means "block who has a voice boos Dilma"), which was generally used in association with **#naovaitercopa**.

Worried about the political scenario that had catalyzed since June 2013, the federal government sought to respond by suggesting a "positive agenda" regarding the World Cup. This scenario spawned the hashtag "the World Cup of the World Cups" (**#CopaDasCopas**), and this phrase was uttered by the president during the draw of the World Cup groups in December 2013. From that point forward, it became a motto for different government agencies and was used by politicians for online pages and profiles.

On January 12, 2014, the Workers' Party (Partido dos Trabalhadores—PT) published the following message on their official Facebook page: "Deal. A good week for everyone who cheers for Brazil. **#VaiTerCopa**." On the same day, the official Facebook profile of President Dilma Rousseff displayed a photo of a fan with the Brazilian team shirt and featured the hashtag **#vaitercopa**.[7] The president's team wrote the message: "CLEAR AND PERFECT—A good week to all who cheer for Brazil." In an interview with *O Estado de S. Paulo* newspaper, the person responsible for managing the PT's social networks (Alberto Cantalice, vice president of the party) stated that the idea "did not have a concrete goal. It's because people demand us to say something."[8]

This official manifestation regarding the ongoing debate in the country, however, triggered a series of reactions on the Internet, which primarily consisted of produced remixes of the photo posted by the presidency that succeeded in undermining the government's legitimacy. Such was the case for the image posted a day later on the "Black Bloc Brasil"[9] page, which featured graffiti of the word "*não*" (Portuguese for "no"), including the anarchist "A" symbol written over the government's original hashtag.[10] In the text, the post stated: "In a burst of boosterism, Dilminha leads us to laugh with the post below (without the intervention of the vandals, of course!). NO TRUCE!"

JOKES AND COMMERCIAL SLOGANS:
OTHER APPROPRIATIONS

We've already said that the typical adoption of hashtags rises and spreads in a bottom-up manner (Bruns & Burgess, 2011). However, even though the typical mode is the most common, it is not the only possibility. In the hashtag war that began with June Journeys, in addition to the markers created by polls in favor and opposed to the FIFA World Cup in Brazil, there have also been variations that point to the use of hashtags in different contexts and with a multiplicity of meanings.

One of these spillovers is the spread of visual memes, particularly in images distributed through Facebook. Designed with playful or ironic characteristics, such memes are shared online and are partially and frequently altered from one person to the next. This is evidenced by variations of the slogan "Keep Calm and Carry On," which originated from a poster drawn up by the UK Ministry of Information during World War II.[11]

Thus, the appropriation of memes prior to the creation of new ones, typical of meme logic (Bauckhage, 2011; Jenkins, 2013), figured in the hashtag war of **#naovaitercopa** vs. **#vaitercopa**. The comedic, ironic character has inspired other appropriations of the hashtag **#naovaitercopa**. In Portuguese, the word for "cup" can refer to any of the following: tree canopy, which is one way of referring to the neighborhood of Copacabana in Rio de Janeiro, the suit of hearts in the card deck, and the room intended for light meals, among others. From the semantic richness[12] of this word, visual memes that played on the hashtag **#naovaitercopa** were elaborated, attaching new meanings to the expression.[13]

In the context of the FIFA World Cup, the adoption of hashtags went beyond the playful applications and was also appropriated by advertising campaigns. Before being touted on the streets and throughout online social networks, for example, the expression "*vem pra rua*" was part of an automobile maker's TV commercial created for the Confederations Cup. On TV and radio, the commercial's jingle invited Brazilians to "come to the streets because the streets are actually Brazil's biggest bleacher [stadium seats]." Protestors also sang this jingle in the streets, illustrating clearly that this intermedia articulation between commercial mass media, social media, fashion, and street art can be seen as an example of contemporary "convergence culture" (Jenkins, 2006).

On the eve of the FIFA World Cup, at least 8 of the 20 partners, sponsors, and official supporters of the event used hashtags on their websites and in their advertising campaigns.[14] Let us consider the hashtags used in campaigns from FIFA supporter Itaú Bank, **#issomudaojogo** ("it changes the game"), and FIFA partner Coca-Cola, **#todomundo** ("everybody"). Below, we will explain these choices and how they were influenced by the proximity to June Journeys.

The **#issomudaojogo** campaign of Itaú Bank, the largest private bank in Latin America, focuses on the influence the Brazilian team supporters would have on the team's results and how Brazilians could get involved with the FIFA World Cup. The campaign that launched in December 2013 points to the influence of the June Journeys: it shows people in the streets, just as they are shown in the demonstrations, but instead of protesting, they are shown cheering for the Brazil national football team. It also includes in its semantic constitution the idea of change, which is highly valued in posters, demands, and discussions regarding the political demonstrations.

Itaú's campaign unfolded with the **#issomudaomundo** hashtag, which relied on websites where citizens' initiatives in the areas of culture, education, sports, and urban mobility were gathered.[15] Again, through these chosen themes, the bank referenced very present characteristics of June Journeys—in addition to its beginning related to urban mobility, posters were common in protests that demanded "standard FIFA" hospitals and schools, in an allusion to the high investment made in stadiums.

Perhaps at the time the campaign was launched, great reactions were not visibly evident in the streets—although many bank branches had their facades broken during the protests of June Journeys. The popular reaction to the advertising campaign differed with Coca-Cola. Launched in February 2013, the Coca-Cola campaign, marked by the hashtag **#todomundo**, claimed that the World Cup would be for all Brazilians. During June Journeys, the idea was poorly received. Billboards and panels were burned in several cities around the country.[16] There were also urban interventions on the billboards, modifying their meaning and erasing the hashtag **#todomundo**.[17]

The reaction can be seen as part of a symbolic dispute of the type seen often on the Internet (Lindgren & Lundström, 2011). However, while the dispute around hashtags on Twitter generates battles for trending topics and discussions about the appropriate use of markers, the dispute around hashtags in the streets displays a physical expression. The spillovers that allow for the adoption of hashtags beyond digital networks also mark the encounter with other logistics that are incompatible with digital ties.

CONCLUSIONS

The collective use of hashtags in Brazil that relate to sport events such as the 2013 FIFA Confederations Cup and 2014 FIFA World Cup showed the political entanglement of the street and social media in contexts of social mobilization. The hashtags engineer intermedia connections around similar or divergent themes, and connect political positions in the porous interfaces between the streets and social media, as with the exemplary case of **#vemprarua**.

The cohesions forming around hashtags are varied and circumstantial. Instead of a common speech organized hierarchically by representative bodies such as political parties and social movements, the wars of hashtags make visible the heterogeneous, circumstantial, and even contradictory behavior that conforms to political assemblages that are typically contemporaries (Latour, 2003). Even with the seemingly polarized disputes centered on the terms **#nãovaitercopa** and **#vaitercopa**, a large variation is identified in the ad hoc communities formed around hashtags and in the meanings assigned to these expressions.

Intermedia connections around hashtags expand their social reach and promote remixes of the typical meme logic (Bauckhage, 2011; Jenkins, 2013). There are, in this context, overlaps among advertising, government, and citizens' speech which not only strengthen the political relevance of hashtags but also evidence their prominence in contemporary socio-communicational dynamics. The overlapping political semantics surrounding advertising hashtags, which in turn have appropriated hashtags disseminated by the 2013 demonstrators, emphasize the increasingly intense political dynamics of sharing that integrate social media with the streets of Brazil. This is very important because it underlines the political complexity of hashtag use—how the unifying power of hashtags is evidenced not only in their political and social use but also in their economic uses, as in the case of advertising campaigns of Itaú Bank and Coca-Cola. The large-scale deployment of hashtags related to the World Cup in various contexts shows that dynamic intermedia is a major parameter of contemporary communicational logic. Nowadays, the character # followed by words not only works as an aggregation resource in social media but also announces themes and slogans in mass media and urban spaces.

NOTES

1. The authors thank CNPq, Fapemig, and PRPq/UFMG for financing the research projects that helped generate this article. Correspondence concerning this article should be addressed to Departamento de Comunicação Social/FAFICH—Avenida Presidente Antônio Carlos, 6627 – Pampulha, Belo Horizonte—MG. CEP 31270-901. Brazil.

2. For more information on the June Journeys, also called the "The Vinegar Revolt," see http://globalvoicesonline.org/specialcoverage/brazils-vinegar-revolt/

3. Examples of urban interventions with hashtags during these protests may be found at http://www.pinterest.com/joanaziller/naovaitercopa/

4. In Brazil, the inclusion of such terms on walls and other urban facilities leads to divided opinions. Part of the population sees it as vandalism to paint upon something that belongs to the public. Another part understands this type of urban inscription as legitimate political expression.

5. https://www.facebook.com/copacbh/posts/599948340092138

6. See, for example, this article published in March 2014: http://www.reuters.com/article/2014/03/28/us-brazil-politics-rousseff-idUSBREA2Q17Y20140328

7. https://www.facebook.com/photo.php?fbid=607535742633354&set=a.351365628250368.87876.351338968253034&type=1&theater

8. http://blogs.estadao.com.br/radar-politico/2014/01/13/pt-responde-ao-movimento-naovaiter-copa-e-cria-o-vaitercopa/

9. During and after June Journeys, almost all the violent actions against the FIFA World Cup and its sponsors were blamed on the Black Blocs arrangements.

10. https://www.facebook.com/photo.php?fbid=571342679625396&set=a.2336013933-99528.53040.232857860140548&type=1&stream_ref=10

11. Examples of the Keep Calm meme can be seen in the gallery of images available at http://www.pinterest.com/alfredovela/keep-calm/. There is even a meme generator, the Keep Calm-o-matic, which also explains the origin of the use of "Keep Calm" (http://www.keepcalm-o-matic.co.uk/)

12. The Houaiss dictionary, one of the most respected dictionaries of the Portuguese language, lists 14 definitions for the word "cup."

13. Some examples of the use of the hashtag #naovaitercopa with extended meaning can be seen at http://youpix.virgula.uol.com.br/memepedia/memepedia-nao-vai-ter-copa/

14. Among the partners, Coca-Cola and Adidas use hashtags on their websites; among the sponsors, Budweiser, Castrol, Oi, and MayPark; and among the supporters, Liberty and Itaú.

15. The campaign website is http://www.issomudaomundo.com.br

16. Images are available at http://www.meioemensagem.com.br/home/comunicacao/noticias/2013/07/18/Coca-Cola-usa-protestos-em-comercial.html and http://img33.imageshack.us/img33/3643/o64y.jpg

17. Images are available at http://rebaixada.org/zoeiros-brincaram-nossa-realidade-issococa-nos-aguarde-welcome-t/, http://rebaixada.org/welcome-to-the-real-world-cup-2014bem-vindos-copa-do-mundo-real-httprealw/

REFERENCES

Bastos, M. T., Recuero, R., & Zago, G. (2014). Taking tweets to the streets: A spatial analysis of the Vinegar Protests in Brazil. *First Monday, 19*(3). Retrieved from http://firstmonday.org/ojs/index.php/fm/article/view/5227

Bauckhage, C. (2011). Insights into Internet memes. *Proceedings of the Fifth International AAAI Conference on Weblogs and Social Media.* Retrieved from http://www.aaai.org/ocs/index.php/ICWSM/ICWSM11/paper/download/2757/3304

Bennett, W. L., & Segerberg, A. (2012). The logic of connective action. *Information, Communication & Society, 15*(5), 739–768. Retrieved from http://dx.doi.org/10.1080/1369118X.2012.670661

Bruns, A., & Burgess, J. (2011). *The use of Twitter hashtags in the formation of ad hoc publics.* Paper presented at 6th European Consortium for Political Research General Conference, 25–27 August 2011, Reykjavik, Iceland. Retrieved from http://eprints.qut.edu.au/46515

Callon, M. (2006). What does it mean to say that economics is performative? In D. MacKenzie, F. Muniesa, & L. Siu (Eds.), *Do economists make markets? On the performativity of economics* (pp. 310–357). Princeton, NJ: Princeton University Press.

Coenen, C., Hofkirchner, W., & Nafría, J. M. (2012). New ICTs and social media in political protest and social change. *International Review of Information Ethics, 18*(12), 2–8.

Herkman, J. (2012). Convergence or intermediality? Finnish political communication in the new media. *Convergence: The International Journal of Research into New Media Technologies, 18*(4), 369–384.

Jenkins, H. (2006). *Culture of convergence: Where old and new media collide.* New York: NYU Press.

Jenkins, H. (2013). *Spreadable media: Creating value and meaning in a networked culture.* New York: NYU Press.

Kastrup, V. (2004). A rede: Uma figura empírica da ontologia do presente. In A. Parente (Ed.), *Tramas da rede: Novas dimensões filosóficas, estéticas e políticas da comunicação* (pp. 80–90). Porto Alegre, Brazil: Sulina.

Latour, B. (2003). What if we talked politics a little? *Contemporary Political Theory, 2,* 143–164.

Latour, B. (2005). *Reassembling the social—An introduction to actor-network-theory.* Oxford, UK: Oxford University Press.

Lindgren, S., & Lundström R. (2011). Pirate culture and hacktivist mobilization: The cultural and social protocols of #WikiLeaks on Twitter. *New Media & Society, 13*(6), 999–1018.

Rancière, J. (2008). *The politics of aesthetics: The distribution of the sensible.* London & New York: Continuum.

CHAPTER NINE

#FuckProp8: How Temporary Virtual Communities around Politics and Sexuality Pop Up, Come Out, Provide Support, and Taper Off

JENNY UNGBHA KORN

INTRODUCTION

Online communities are supposed to last. In fact, a typical academic would consider a virtual community that has gone inactive as a failed one. Past research on digital communities regularly emphasizes thriving practices and survival strategies for keeping such communities vibrant and lively (Baym, 1999; Fox, 2004; Howard, 2010). In contrast, I take a different approach to online communities that have gone inactive: I interpret the stasis of communities on the Internet as a healthy, logical, and even anticipated stage of event-based online communities. From my perspective, some online communities are supposed to endure only temporarily.

In this chapter, I examine a specific event-based case to illustrate the temporality of online communities that arise, cohere, and then taper into stasis. Through focusing on virtual communities whose longevity is tied to a particular event, I address an underresearched area of online community scholarship, namely, those characterized as temporary. For my case study, I chose opposition to Proposition 8, legislation that banned same-sex marriage in California (Liptak, 2013). I analyze how politics and sexuality are intertwined with discourse and identity through communities organized by Twitter hashtags. I also discuss how these online communities contribute to a counterpublic whose discourses concurrently featured acceptance for nonmainstream sexuality while challenging anti-queer legislation (Milioni, 2009).

COMMUNITY EMBODIMENT MODEL AND DIGITAL PUBLICS

To be clear, angry publics that use hashtags on Twitter want to be found, to be heard, and to be distributed through tweets and retweets (see also Rambukkana, Chapter 2, this volume). In an online rendering of an "unruly practice" in which power, discourse, and sexuality engage publicly through Twitter, Nancy Fraser's (1989) work is reimagined in political engagement via tweets. This study explores a loud counterpublic that is also intentionally temporary. Through Twitter and Facebook, virtual communities formed in opposition to the Proposition 8 ban on same-sex marriage in 2008 and then again in 2010 and 2012, when the proposition was passed, overturned, and the overturn affirmed by courts in California, a decision that was upheld by the United States Supreme Court in 2013 (Cillizza & Sullivan, 2013; Liptak, 2013). The dominant interpretative public for the passing and support of Proposition 8 was anti-queer, but as people exercised supportive agency via online social networks about the impact of this proposition, a counterpublic of queer empowerment emerged. In the case of Proposition 8, online counterpublics were useful in negotiations about sexuality, marriage, and love, issues that are both private and public in nature (Fraser, 1989; Habermas, 1991; Milioni, 2009; Papacharissi, 2010).

Within this chapter, I employ a theoretical framework in the community embodiment model (CEM) to examine community embodiment of anti–Proposition 8 online communities "as a melding of the physical and the virtual that are embodied by the imagined" (Fox, 2004, p. 48). This "doubling of space" of the physical and the virtual accurately describes Proposition 8 as both a physical event through in-person rallies and protests across the country and as a mediated event through online communities connected through Facebook and Twitter (Meyrowitz, 1985). Individuals who participated in online communities that opposed the passage of Proposition 8 formed a counterpublic to the dominant narrative of denying marriage to same-sex partners. The collective identity of Twitter and Facebook communities against Proposition 8 reflected broad acceptance of individuals with diverse sexualities and, more specifically, advocated for the choice of marriage for those people of nonmainstream sexual orientations. The case of Proposition 8 illustrates how communities and counterpublics mark and are marked by the Internet through "likes," favorites, retweets, and hashtags.

Using the case of Proposition 8 online communities, I extend the community embodiment model by developing three phases of community embodiment:

formation, impact, and stasis (or current inactivity). These three stages of community embodiment do not occur in discrete moments of time in a linear fashion but rather as general phases through which community occurs. I complicate community embodiment by examining a community that is event based, which leads to temporary community. Communities within the community embodiment model usually work to sustain themselves, but in the case of Proposition 8, the temporal span of online communities working to defeat Proposition 8 is short, leading to an expected, lasting, and current stasis period of CEM after the event has passed (Fox, 2004).

CRITICAL DISCOURSE ANALYSES OF HASHTAG PUBLICS

In this study, I utilize critical discourse analyses to examine online social support communities that form, cohere, and fade based on a specific event. Discourses are social interactions embedded within text and language by actors (Van Dijk, 1985). Anti–Proposition 8 discourse became a means for enacting community online. The injustice around Proposition 8 caused a moment of unity for otherwise unfamiliar people, brought together through event-based hashtags, including **#FuckProp8** and **#RejectProp8**. Digital discourses on Facebook and Twitter covered the time period of 2008 through 2014. Tweets that used hashtags with the word "Prop8" and Facebook pages that used "Prop 8" or "Proposition 8" were included in the study. Specifically, the dataset presented in this book chapter on hashtag publics focused on community discourse surrounding hashtag users of **#FuckProp8** and **#RejectProp8**.[1] Choices in language render power; creation of specific hashtags influences community boundaries. Critical discourse analyses view language as a form of social practice and focus on the ways social and political domination are reproduced in text and talk (Fairclough, 1995; Van Dijk, 1993).[2] I apply critical discourse analyses to the data with particular attention paid to linguistic dynamics across different social groups. Through critical discourse analyses, I examine how a queer-accepting public manifests in opposition to dominant discourse found within an anti-queer public and how digitally networked individuals create counternarratives to build community online against Proposition 8.

HASHTAG PUBLIC FORMATION

Under the community embodiment model, community formation may be viewed as the intentional act of connection that an individual chooses to join online conversations by others. Specifically, choosing to use a hashtag in one's tweets is to mark the tweet as of interest for not just the potential audience of only one's (usually small number of) followers but also the possible readership of the much larger portion of the Twitterverse that pays attention to that hashtag. To use a hashtag is to engage in the wider discussion; it ties one's tweet to the community of tweets on the same topic (Johnson & Korn, 2013).

Created in 2008, Proposition 8 was a proposition to the state of California's constitution that would oppose same-sex marriage. The wording of Proposition 8 was (intentionally, some have said) tricky: to be in support of Proposition 8 was to become part of an anti-queer public. Voters in favor of Proposition 8 declared that California should ban gay marriage in the future (Cillizza & Sullivan, 2013; Liptak, 2013). However, the passage of Proposition 8 in 2008 also gave birth to a different public online. In 2008, communities that formed on Facebook, Twitter, and other websites constructed a counterpublic advocating same-sex marriage and queer acceptance. On October 6, 2008, "No on Proposition 8" formed as a page on Facebook. On Twitter, hashtag names proclaimed their purpose: #FuckProp8 and #RejectProp8. These communities were united upon a single, concrete action: defeating Proposition 8.

Counterpublics interact with the dominant population (Habermas, 1991; Milioni, 2009). To combat a public that supported the passage of Proposition 8, a counterpublic of queer acceptance arose. Individuals voiced their opposition to Proposition 8 through founding and contributing to Internet communities whose online users exercised public agency through the use of hashtags, "likes," and forwards. As certain hashtags gained validity through frequency of use, communities manifested around queer acceptance. Queering, or departures from normative practices, of online communities by queer individuals (among others) appeared as the use of intentionally inflammatory words, including "fuck." A tweet that contains only #FuckProp8 signifies the user's participation in community development against the passage of Proposition 8. The tweet may not reveal an explanation for the stance, but it contributes to a growing counterpublic of users that support same-sex marriage. The hashtag itself by its title signifies rejection of Proposition 8, and its implementation amplifies the power of the counterpublic. Queer modalities of expression online subvert normative ideas of community leadership and boundaries. Through Twitter, online users disrupted traditionally hierarchical structures of leadership within communities; participation in formation of hashtag communities did not require administrative approval or membership vetting.

In the formation of support communities, unlike traditional in-person support groups, personal experience with the triggering event may or may not be necessary for participation. The support required is usually sought from those who have experienced similar life situations. However, for many individuals, such community, especially around lesbian, gay, bisexual, transgendered/sexual, queer/questioning, and intersex (LGBTQI) issues that often contain social stigma, cannot be achieved through their primary relationships. Joshua Meyrowitz's (1985) concept of the "doubling of space" explains that events occur in both a physical space and a mediated one. Particularly for Twitter users, the doubling of space is important, as members of the community may experience the initial media event apart from one another but can experience it together in the mediated space created online. Though Proposition 8 directly affected same-sex couples in California, members in anti–Proposition 8 online communities spanned sexual and geographical boundaries, united in their search for support, information, and activism.

The choice to go online enabled an increase in membership in a queer-accepting public that was not necessarily constrained by geographical location. While the event that drew individuals was the fight for the defeat of Proposition 8, the community coalesced around the belief that marriage should be open to people of all sexual orientations. This online community went beyond the boundaries of the issue of the state constitutional amendment in California to include those on the Internet who shared an orientation towards social justice that included marriage equality for everyone. Tweets that announced protests, rallies, and petitions against Proposition 8 were not confined to California, because political gatherings were occurring across the United States in support of marriage equality. Tweet discourse highlighted how the legalization of same-sex marriage affected not just those seeking matrimony but also the families of those individuals, particularly adult children who supported the legal unions of their same-sex parents.

This hashtag public offered the opportunity to reclaim and reconstruct a different public, one in which LGBTQI identities are accepted and affirmed. From the grassroots up, an alternative narrative was configured through online community formation around opposition of Proposition 8. Within Tweet discourse, support for queer acceptance mapped onto support for same-sex marriage: individuals within the **#RejectProp8** and **#FuckProp8** communities often conflated anti–Proposition 8 sentiment with general LGBTQI support. For example, a major theme that emerged from the anti–Proposition 8 discourse was equality of love. "Love is love" was repeated throughout **#RejectProp8** tweets as an assertion that the love between heterosexual couples and the love between queer couples should be considered equal, socially and legally. A variant of that theme was that "love is universal."

HASHTAG PUBLIC IMPACT

Injecting sexual orientation into the Twitterverse reminds Twitter users that sexuality is present online, even when it is not explicitly referenced. Gender, race, sexuality, class, and other components of cultural identity tend to be overlooked as those in dominant populations exercise positions of privilege in their ability not to think about their linkages to the majority. By drawing attention to the existence of alternative sexual preferences, the counterpublic of queer acceptance marks and is marked by Twitter as a space that is explicitly sexual and political, highlighting subjectivity and power. This explicit focus on sexuality is an important impact of an electronic counterpublic. The Proposition 8 case demonstrated the transformation of Twitter into a place where individuals may create awareness of the existing dominant public by establishing an alternative public that attacks the dominant narrative (Rocco & Gallagher, 2006; Ryder, 1991; Tollefson, 2010).

Using the Internet to galvanize an anti–Proposition 8 counterpublic, individuals who participated in Facebook- and Twitter-based communities helped to construct a loud sphere of queer acceptance and support. The Internet offered a means for creating a counterpublic that addressed, interrogated, complicated, and unpacked the dominant narrative. The impact of online communities went beyond the activation of online content. While attention often centralizes around the creation of hashtags and pages, the effect of online communities may also be felt by lurkers and readers. In other words, both vocal and quiet participants aided in the construction of an online counterpublic. Posts within loud counterpublics inform online users that other voices besides mainstream narratives exist. Community members who followed a hashtag or page may have chosen never to "like," favorite, or retweet a post; nevertheless, they were a vital part of an online community that is easily overlooked, and hard to capture due to their silence. With Twitter's new "check out your week on Twitter" function, Twitter users are informed of the number of people who view their accounts and the number of individuals who read each individual tweet, even if those visitors do not leave a "favorite" or retweet. Lurker opinions may be influenced by the online discussion generated within the counterpublic, and lurkers may even eventually be encouraged to contribute to online communities via posts (Ridings, Gefen, & Arinze, 2006; Takahashi, Fujimoto, & Yamasaki, 2003).

An impact of online community is the ability to be active in digital spaces that are present but that the body is unable to attend. Celebrity chef Hugh Acheson may not have been in California, but the Canadian chef who now considers Georgia his home joined the anti–Proposition 8 community when he tweeted, "I honestly feel like this country is just making more progress than ever. #FuckProp8" on March 26, 2013. His tweet indicated his support of the overturn of Proposition 8, while explicitly referencing the effect of Proposition 8 as national, not local to California. His tweet was also representative of how legislation on same-sex marriage affects all citizens, straight and LGBTQI. Some celebrities joined the counterpublic of opposition to Proposition 8 through adding their own tweets to existing hashtags, but others, such as Drew Barrymore, were brought into the conversation by individuals who had read supportive celebrity tweets and then retweeted them with the appropriate hashtag. Businesses such as Levi Strauss also contributed to a counterpublic of gay acceptance through public Tweets attached to anti–Proposition 8 hashtags, capitalizing upon queer-friendliness as socially just and economically profitable ("Levi Strauss," 2008).

Online communities may be particularly important for marginalized groups of people. In previous research, I have found that LGBTQI individuals use Facebook groups to find others like them, particularly in Asia (Korn, in press). In creating and finding community for queer Asians, Asian males use Facebook groups to elide the stigma often attached to cultural understandings of homosexuality (Poon & Ho, 2008). Stigma surrounding queerness encourages LGBTQI individuals to find online communities rather than physical ones.

Because of Proposition 8, outness manifested as content and as process. "Out" holds special significance as cultural knowledge within the LGBTQI community (e.g., see Sedgwick, 1990). Tweets that employed the word "out" played upon the simultaneity of common usage and specialized jargon for queer individuals. For example, discourse around "standing OUT" could be read as promoting uniqueness in individuals who attend protests physically, perhaps in their apparel choice. But a more nuanced interpretation views such tweets as encouragement to come out of hiding within the heterosexual closet and embrace queer sexuality openly.

The media event of Proposition 8 became a catalyst for individuals, including celebrities, to come out themselves: in response to Proposition 8, comedian Wanda Sykes chose to come out as gay, helping to create a counterpublic based on identity formation. While giving an angry, impromptu speech at an anti–Proposition 8 rally in Las Vegas, Nevada, in 2008, Sykes came out publicly and emphasized that her marriage to her wife, one among the 18,000 LGBTQI legal marriages in California, was at stake (McKinley, 2009; McLaughlin, 2011; Sykes, 2008). Coming out, or disclosing sexual orientation to others, is a lifelong process in which decisions to reveal one's sexual identity occur episodically. Every time lesbian, gay, bisexual, transgendered, queer, and intersex people voluntarily share about their sexuality, they engage in the iterative process of coming out as accepted and integrated (Coleman, 1982; Levesque, 2012). Utilizing online social media in the coming-out process enables individuals to identify as LGBTQI publicly and to find others who have experienced similar life situations.

HASHTAG PUBLIC STASIS

The intensity of the formative and impact stages of these improvisational online communities faded with the passage of time. The anti–Proposition 8 public lasted as long as Proposition 8 remained a threat to the gay-acceptance public. In June 2013, when the United States Supreme Court declined to decide the Proposition 8 case, California became the thirteenth state in the country to allow same-sex marriage (Liptak, 2013).[3] Discourse in 2013 from **#FuckProp8** celebrated the success of the defeat of Proposition 8 and emphasized how "today is a good day," "same sex marriage is legal in California," and how "marriage is just marriage and how love is love." Similarly, **#RejectProp8** discourse was active in 2008 and 2009, but it steadily declined after 2010. Since then, communities surrounding both hashtags have been in stasis.

According to conventional research, communities are relevant while they thrive (Baym, 1999; Fox, 2004; Howard, 2010). Communities in stasis are a frequent topic in the biological sciences but not in the social sciences. Yet stasis for event-based communities represents an important stage for Twitter hashtag publics. Individuals in anti–Proposition 8 online communities marked the end of years of activism against Proposition 8 with tweets that reaffirmed the country's move towards progressive politics, signaling a shift in public opinion about the acceptance of same-sex marriage. The success of activism against Proposition 8 coincided with the cessation of its event-based hashtag communities. With nearly 255 million monthly active Twitter users (Twitter, 2014), the propensity exists for a revival of either **#RejectProp8** or **#FuckProp8**, but members of both communities have not chosen to tweet with either hashtag. The inactivity of anti–Proposition 8 hashtag publics represents a natural conclusion to the end of the centralizing force behind community creation. Such stasis is not a failure but rather a sign of normality in the evolution of an event-based community online.

LIMITATIONS AND FUTURE DIRECTIONS

The formation of online communities is constrained structurally by individuals who have access to the Internet, though not necessarily to those with user accounts within Facebook and Twitter. For example, Facebook pages are public, such that online visitors may visit the page for content without creation of a Facebook account. Similarly, those Twitter accounts not set to private have tweets that are publicly accessible. Structural constraints impact individual decisions to join online communities. This study is representative of individuals who have chosen to become members of Twitter and tag their tweets with hashtags against Proposition 8: these individual decisions create larger patterns of event-based online community embodiment.

The findings presented in this chapter focus on conformist behavior within online community formation and development, i.e., pro-queer content that matched the hashtag of **#FuckProp8**. Some hijacking of hashtags also occurs whenever a hashtag becomes popularized. The hacking of such hashtags may range from the benign, e.g., "I love coffee **#RejectProp8**," to the pointed, e.g., "Christians would not support **#FuckProp8**."[4] In a future study, I will focus on online content that was discordant with queer acceptance ideals set by the counterpublic.

An examination of how the dominant narrative attacks and tries to minimize the counterpublic online is also forthcoming in a study of Proposition 8 hashtag communities.

CONCLUSIONS

By choosing to create communities online, individuals contribute to the practice of building solidarity around events in a public way that is mediated by computers. Meaningful experience with the ease of creating and participating in online communities encourages users to include the Internet as a location for finding communality. A form of social entrepreneurship, developing event-based communities online becomes easier with practice. The pattern of activity is not only theorized in scholarship; it is felt in implementation by community participants. An impact of event-based communities online is to construct publics that are counter to the dominant narrative, publics that react, redefine, and reimagine discourse from emergent communities. Going to the Internet is a way of leveling the playing field for individuals active in the production of counterpublics. While the mainstream public is vetted and validated by mainstream public sources, such as television, the counterpublic may form and cohere across online public sources, which may result in mainstream attention as a consequence. A type of cultural conditioning is in progress through the establishment of counterpublics through the Internet: online communities may exist temporarily, but their influences are powerful during their life spans.

"Community," "public," and "politics" are polysemic terms demanding definition and redefinition. They are technological and technologized. In this study, community and public are represented, symbolized, and enacted via hashtags. Proposition 8 offers a snapshot of how online communities may quickly convene around a national event and then eventually taper off as the event reaches conclusion and the desired effect has manifested. Online communities that are temporary are often labeled as failures, but I argue that the temporariness of event-based communities is an expected and healthy stage of event-based community development online. In the case of Proposition 8, the current stasis of online communities reflects success in defeating anti-queer legislation. Multiple, temporary anti–Proposition 8 communities constructed a counterpublic that advocated same-sex marriage, affirmed queerness, encouraged coming out, and provided social support online. This study contributes to a modern understanding of online community formation, impact, and inactivity. Proposition 8 provides an example of emergent discourse communities reflective of today's culture of "searchable talk" (Zappavigna, 2011), one built around hashtags that are angry, loud, powerful, and temporary.

NOTES

1. Facebook information is provided as context only here.
2. Speaking of language, I use "queer" as an umbrella term for all individuals whose sexual orientation is not considered mainstream, which includes communities commonly termed as lesbian, gay, bisexual, bicurious, transsexual, transgendered, queer, and intersex (LGBTQI). I also use "online" interchangeably with "digital" and "virtual" to speak of Internet-mediated communities.
3. California was an early adopter of same-sex marriage, becoming the second state in the United States, after Massachusetts, to allow same-sex marriage, in 2008, briefly, pre–Proposition 8.
4. For a similar banal, rejecting, or mocking use of political hashtags, in this case for backlash against a feminist tag, see Antonakis-Nashif, Chapter 7, this volume.

REFERENCES

Baym, N. K. (1999). *Tune in, log on: Soaps, fandom, and online community* (Vol. 3). Thousand Oaks, CA: Sage.

Cillizza, C., & Sullivan, S. (2013, March 26). How Proposition 8 passed in California—and why it wouldn't today. *The Washington Post.* Retrieved from http://www.washingtonpost.com/blogs/the-fix/wp/2013/03/26/how-proposition-8-passed-in-california-and-why-it-wouldnt-today

Coleman, E. (1982). Developmental stages of the coming out process. *Journal of Homosexuality, 7*(2–3), 31–43.

Fairclough, N. (1995). *Critical discourse analysis: The critical study of language.* Harlow, UK: Longman Group Limited.

Fox, S. (2004). The new imagined community: Identifying and exploring a bidirectional continuum integrating virtual and physical communities through the community embodiment model (CEM). *Journal of Communication Inquiry, 28*(1), 47–62. doi 10.1177/0196859903258315

Fraser, N. (1989). *Unruly practices: Power, discourse, and gender in contemporary social theory.* Cambridge, UK: Polity Press.

Habermas, J. (1991). *The structural transformation of the public sphere: An inquiry into a category of bourgeois society.* Cambridge, MA: MIT Press.

Howard, T. (2010). *Design to thrive: Creating social networks and online communities that last.* Burlington, MA: Morgan Kaufmann.

Johnson, B. (Interviewer), & Korn, J. (Interviewee). (2013). New Miss America gets the crown with some online racism on the side [Interview audio file]. Retrieved from http://www.marketplace.org/topics/tech/new-miss-america-gets-crown-some-online-racism-side

Korn, J. (in press). Black nerds, Asian activists, and Caucasian dogs: Online race-based intercultural group identities within Facebook groups. *International Journal of Intelligent Computing in Science & Technology* (IJICST).

Levesque, R. J. (2012). Coming out process. In R. J. Levesque (Ed.), *Encyclopedia of Adolescence* (pp. 478–479). New York: Springer.

Levi Strauss pairs with PG&E to fight Proposition 8. (2008, September 27). *The Advocate.* Retrieved from http://www.advocate.com/news/2008/09/27/levi-strauss-pairs-pgampe-fight-proposition-8

Liptak, A. (2013, June 26). Supreme Court bolsters gay marriage with two major rulings. *The New York Times*. Retrieved from http://www.nytimes.com/2013/06/27/us/politics/supreme-court-gay-marriage.html

McKinley, J. (2009, May 25). California couples await gay marriage ruling. *The New York Times*. Retrieved from http://www.nytimes.com/2009/05/26/us/26gay.html

McLaughlin, K. (2011, June 3). Wanda Sykes on coming out: 'I kind of shocked myself.' *CNN*. Retrieved from http://www.cnn.com/2011/SHOWBIZ/06/02/piers.morgan.wanda.sykes

Meyrowitz, J. (1985). *No sense of place: The impact of electronic media on social behavior*. New York: Oxford University Press.

Milioni, D. L. (2009). Probing the online counterpublic sphere: The case of Indymedia Athens. *Media, Culture & Society, 31*(3), 409–431.

Papacharissi, Z. (2010). *A private sphere: Democracy in a digital age*. Cambridge, UK: Polity Press.

Poon, M. K.-L., & Ho, P. T.-T. (2008). Negotiating social stigma among gay Asian men. *Sexualities, 11*(1–2), 245–268.

Ridings, C., Gefen, D., & Arinze, B. (2006). Psychological barriers: Lurker and poster motivation and behavior in online communities. *Communications of the Association for Information Systems, 18*(1), 16.

Rocco, T. S., & Gallagher, S. J. (2006). Straight privilege and moral/izing: Issues in career development. *New Directions for Adult and Continuing Education, 2006*(112), 29–39.

Ryder, B. (1991). Straight talk: Male heterosexual privilege. *Queen's Law Journal, 16*, 287–290.

Sedgwick, E. K. (1990). *Epistemology of the closet*. Berkeley: University of California Press.

Sykes, W. (2008). *Stand out for equality* [Video speech]. Retrieved from http://www.youtube.com/watch?v=S6UdoCIYvIw

Takahashi, M., Fujimoto, M., & Yamasaki, N. (2003, November). The active lurker: Influence of an in-house online community on its outside environment. In *Proceedings of the 2003 international ACM SIGGROUP conference on supporting group work* (pp. 1–10). New York: Association for Computing Machinery.

Tollefson, K. (2010). *Straight privilege: Unpacking the (still) invisible knapsack*. Retrieved from http://files.eric.ed.gov/fulltext/ED509465.pdf

Twitter. (2014). *About: Company*. Retrieved from https://about.twitter.com/company

Van Dijk, T. A. (1985). Introduction: Levels and dimensions of discourse analysis. In T. A. Van Dijk (Ed.), *Handbook of discourse analysis* (pp. 1–11). London: Academic Press.

Van Dijk, T. A. (1993). Principles of critical discourse analysis. *Discourse Society, 42*, 249–283.

Zappavigna, M. (2011). Ambient affiliation: A linguistic perspective on Twitter. *New Media & Society, 13*(5), 788–806.

CHAPTER TEN

More than Words: Technical Activist Actions in #CISPA

STACY BLASIOLA, YOONMO SANG, AND WEIAI WAYNE XU

In recent years, Internet policy making has received an increased level of interest from publics (Breindl & Briatte, 2013; Powell, 2012). The net neutrality debate and the SOPA/PIPA bills, for example, both relate to Internet policy making, and each case has garnered a fair share of the spotlight. Understanding whether and how Internet activism can affect policy-making processes is an issue that warrants our attention. Previous studies have investigated digital activists' communicative and technical tactics (Van Laer & Van Aelst, 2010); however, most studies fail to provide a workable investigation into the range of technical actions used and how those actions utilize the affordances of the Internet.

To that end, this study examines how opponents of the Cyber Intelligence Sharing and Protection Act (CISPA) bill have used the **#CISPA** Twitter hashtag to employ both communicative and technical actions. It situates anti-CISPA activists as operating within the context of online social movements and categorizes their tactical behaviors as technical activist actions (TAA) (Powell, 2011). Building upon the concept of TAA and previous studies on repertoires of online social movements (Costanza-Chock, 2003; Van Laer & Van Aelst, 2010), this study refines and differentiates communicative from technical actions in digital activism.

EMPOWERMENT OR SLACKTIVISM?

Scholars are divergent on whether information and communication technologies (ICTs) can indeed change political courses and outcomes (Earl & Kimport, 2011). One group of scholars claims that using ICTs for protest purposes "primarily increases the size, speed, and reach" of communications but has little effect "on the processes underlying activism" (Earl & Kimport, 2011, p. 24). They note that online activism tends to entail neither high-level efforts, risk taking, nor offline collective action. Using the term "slacktivism," Morozov (2009) argued that virtual participation is low cost, low impact, and only leads to users' personal satisfaction. Moreover, the existence of the digital divide and the difficulty of generating and maintaining stable ties among activists may also hinder the outcome of online collective actions (Van Laer & Van Aelst, 2010).

In contrast, a group of scholars have pointed out that ICTs enable new forms of collective action with varying degrees of efforts and consequence (Earl & Kimport, 2011; McAdam, Tarrow, & Tilly, 2001; Van Laer & Van Aelst, 2010). Van Laer and Val Aelst (2010) developed a typology of new digitalized collective actions: (1) Internet supported and low threshold, (2) Internet supported and high threshold, (3) Internet based and low threshold, and (4) Internet based and high threshold. Internet-based action refers to an action performed solely online, while Internet-supported action refers to an offline activity that can be facilitated and made easier by the Internet to organize and manage. Low- and high-threshold actions reflect a "hierarchy of political participation" (Van Laer & Van Aelst, 2010, p. 1150). Some actions require more time investment, commitment and sophisticated technical skills and could invoke legal and political consequences (Van Laer & Van Aelst, 2010).

TECHNICAL ACTIVIST ACTIONS

Building upon the aforementioned typology, this case study asks to what extent digital activists' communicative actions on Twitter function as technical actions. In so doing, we use the framework of technical activist actions (TAA) defined by Powell (2011) as those that use or exploit the existing structure of the Internet to challenge institutions of power. Various technical activities range from massively producing messages, which requires little technological skill, to technologically sophisticated ones such as carrying out Internet blackouts, DDoS attacks, and document or information releases (Powell, 2011, 2012).

Twitter functions both as an interpersonal tool that enables point-to-point delivery of messages and as a mass media platform which average users subscribe to and use to broadcast other-provided information (Kwak, Lee, Park, & Moon,

2010). When used for discussions of public affairs, Twitter users employ hashtags. Hashtags are words or phrases prefixed with the symbol #, used for categorizing tweets pertaining to the same topics. Hashtag use fosters imagined communities based on common interest (Bruns & Burgess 2011; Gruzd, Wellman, & Takhteyev, 2011). When the public interest is involved and communities spring up around hashtags, they are referred to as ad hoc or hashtag publics (Bruns & Burgess, 2011, and Chapter 1, this volume).

The role of Twitter in social movements has been examined in communication contexts. Previous studies examined the role of non-elites in disseminating information on Twitter (Hermida, Lewis, & Zamith, 2014; Lotan et al., 2011), the usage of frames in tweets (Meraz & Papacharissi, 2013), and the news values that emanated in tweets (Papacharissi & de Fatima Oliviera, 2011). Twitter messages in activism entail varying degrees of engagement as described in Lovejoy and Saxton's (2012) information-community-action framework. For example, purely informational tweets are devoid of emotion and opinion and are less engaging than tweets that provide opinion and enable social interaction (Xu, Sang, Blasiola, & Park, 2014). Community tweets are described as those that serve to build community (Lovejoy & Saxton, 2012; Xu et al., 2014). The most engaging of all are action tweets—calls to action that encourage users to participate in concrete actions such as signing petitions, making a donation, joining protests, etc. (Xu et al., 2014).

CISPA OVERVIEW

The Cyber Intelligence Sharing and Protection Act was intended to assist cybersecurity by allowing "federal agencies and private companies to share customer information" (Chen, 2012, para 2). CISPA has caused controversy because it would bypass warrants and allow companies to share information "with any other entity, including the federal government" (McCullagh, 2013, para 4). Despite originally failing in the Senate in 2012, CISPA re-emerged after President Obama issued an executive order in February 2013 calling for a comprehensive cybersecurity bill. Although CISPA passed the House vote a second time in April 2013, it was ultimately shelved by the Senate for good. As CISPA debates were heard on Capitol Hill, opponents of CISPA took to Twitter, where they organized efforts to battle the bill.

The present study entails three goals. First, it investigates broadly what type of content is presented in #CISPA tweets using the aforementioned information-community-action framework. It then evaluates tweets to determine whether they contain TAAs. Lastly, the study discusses the source characteristics of URLs included in tweets. Thus, we ask the following research questions:

RQ1: Based on the information-community-action framework, what content categories are salient in **#CISPA** tweets?

RQ2: Based on the Internet and social movement action repertoires, what TAAs are evident in **#CISPA** tweets?

RQ3: What are the sources in the included URLs?

METHOD

We conducted a content analysis of 1,877 tweets containing the **#CISPA** hashtag. Two coders were used, and intercoder reliability was conducted using Holsti's formula.[1] **#CISPA** was selected because it served as the hashtag used by both proponents and opponents of the bill. Tweets were collected on 20 days between March 21, 2013, and April 10, 2013, using the Twitter Archiving Google Spreadsheet (Hawksey, 2013). The timeframe covers the period after CISPA was reintroduced to the House but before the House vote had taken place. This timeframe captured the efforts of both sides to make their respective cases for or against the bill.

A total of 787 unique users contributed to the 1,877 tweets in the dataset. Sixty-three percent, or 1,185 tweets, contained an @mention. Comparatively, only 27%, or 511 tweets, contained a retweet (RT). Only 6%, or 133 tweets, contained an @reply. The majority of tweets, 1,236 or 66%, contained links to webpages.

RESULTS

To investigate the communicative aspect of the tweets, this study adopts the coding scheme developed by Lovejoy and Saxton (2012). This scheme places tweets in one of three communication categories, Informational, Community, or Action, and each category reflects a correspondingly higher level of engagement. This categorization scheme codes the communicative aspect of the tweet itself. That is, the scheme describes the type of message contained in the tweet. The intercoder reliability coefficient was 88.1%, which indicates an acceptable level of agreement.

Information

The purpose of informational tweets "is solely to inform" (Lovejoy & Saxton, 2012, p. 343). We found several subcategories of tweets that fill the Information function.

Benefactors. A small percentage (5%) of tweets contained information about who would benefit if CISPA passed. These tweets informed people that congressional proponents of CISPA were financially supported by the companies that would benefit from its passing.

Table 1. Tweet Functions.

Category	Example	Frequency	(%)
Information (19%)			
Benefactors	@RepMikeRogers I guess when #CISPA lobbyist are paying you $214,750 in contributions you'll say almost anything. https://t.co/u8yBeUOW5O	92	4.9
Educational	.@freedomhousedc Urges US Congress to Amend Bill That Will Harm Internet #Privacy http://t.co/ftnOBm2w8V #netfreedom #CISPA	210	11.2
Pro CISPA	I realize many of you think software piracy is "cool." It isn't. Piracy costs film studios billions every year. Grow up and support #CISPA.	47	2.5
Community (48%)			
Injustice	@RepGoodlatte @politico @TheJusticeDept Check out the DOJ ignoring the 4th Amt and criminalize and spy on citizens. Repeal #CFAA #CISPA etc	163	8.7
Support	@Infographitweet #S2 @FastCompany video opposing #CISPA - @alexisohanian calls @google Larry Page hilarious results. http://t.co/5R17N-aLTGj	691	36.8
Action (36%)	I've joined 50,000 others in asking @BarackObama to veto #CISPA (the cyber snooping bill). Add your name: http://t.co/ZdI1xofXbx	674	35.9
	Total	1,877	100

Educational. A larger group of Information tweets (11%) provided breakdowns of the bill and explained what it would do. Primarily, Educational tweets had in common the notion that learning about CISPA would make one against it.

Pro-CISPA. Although they represented a minor portion of the dataset, 3% of the tweets supported CISPA and contained information that explained what benefits would come from its passing.

Community

We coded tweets as Community if it appeared that their primary purpose involved social interaction or spreading information aimed at building solidarity (Lovejoy & Saxton, 2012). Under the Community function, we found two subcategories emerge: Injustice and Support.

Injustice. Injustice tweets invoked a sense of outrage and made salient the manners in which CISPA would revoke rights, infringe on privacy, and/or contribute to the government's ability to monitor citizens' online behavior. The Injustice category was present in 9% of the #CISPA tweets.

Support. Tweets that made salient the number of people or companies who supported the fight against #CISPA were coded as belonging in the Support subcategory. These tweets helped to build community because they simultaneously expressed the size of the anti-CISPA community while at the same time encouraging others to take the same position. Support was present in 37% of the #CISPA tweets.

Action

Action tweets are considered the most engaging. In Action tweets, "Twitter users are seen as a resource that can be mobilized" (Lovejoy & Saxton, 2012, p. 345). In the #CISPA discussion, tweets that assume a position against CISPA by calling for or presenting steps for activism were coded as Action tweets. The Action frame was the second-most-used frame in the #CISPA tweets, and was present in 35.9% of the observations.

Technical Activist Actions

Tweets were coded to indicate whether a TAA was present. If so, we coded for which type of TAA the tweet contained. The action coded for was present in either the tweet or in the link the tweet contained. Tweets that did not contain a TAA were coded as purely Communicative. The intercoder reliability coefficient was acceptable at 88.66%. TAAa were found in 1,580 tweets, or 84.2% of the sample. Only TAAs that occurred a minimum of three times are reported below.

Table 2. Technical Activist Actions.

Category	Example	Frequency	(%)
Technical Activist Actions (84%)			
DDoS	Homeland Security Secretary: U.S. Financial Institutions Under Hacker Assault http://ow.ly/1PhUNX #CISPA #DDoS #TPP #SOPA #InternetFreedom	3	.16
Organizing	RT @4thAnon: #Anonymous Message #OpNoShow http://t.co/h0KwubgprC Boycott Hollywood Movies on opening #SixStrikes #SOPA #PIPA #CISPA	9	.48
Mention	Another opponent of #CISPA: Astarte' R. of OR @RepMikeRogers	550	29.3
Information Release	Twoops: #CISPA co-sponsor deletes tweet highlighting his financial ties to pro-CISPA companies: http://snlg.ht/105gzIS.	20	1.07
Petitioning	Now is a good time to sign the ACLU petition against #CISPA, the yet-another-ridiculous-cyber-snooping bill: http://t.co/kR4tFDIvG3	988	52.63
Purely Communicative (16%)			
	How much FAIL is our government when we have to STOP them every year from passing something people don't want. #demonocracy #cispa	307	16.36
	Total	1,877	100

DDoS. This category included tweets that reported, referenced, or called for DDoS attacks or tweets that contained information about how to perform a DDoS attack. Although DDoS attacks are frequently reported by news organizations when they occur, and thus may appear as a common tactic, they represented only .02% of the sample and occurred three times.

Organizing. Tweets that attempted to organize offline efforts to meet up, boycott, or rally were coded as Organizing tweets. They were determined TAAs because they relied on both the RT and social networking function of Twitter to perpetuate the message. Organizing tweets were present in only .4% of the sample.

Mention. Tweets that utilized the @mention function of Twitter to hail political figures were coded as Mention. Frequently, these tweets shared the characteristic of being part of a scheme to send multiple messages to prominent CISPA-related political figures.[2] An example of this is "Another opponent of #CISPA: Astarte' R. of OR @RepMikeRogers." In this tweet, Congressman Mike Rogers, the chairman of the House Intelligence Committee, is hailed in a SPAM-type manner, the sheer volume of which was intended to show the congressman the weight of CISPA's opponents. Mentions were included in 550 tweets in the sample, or 29%.

Petitioning. Tweets that contained access to online petitions were coded as Petitioning tweets. These took advantage of the replicability and scalability of the Internet to perpetuate the message and amass large numbers of petition signers. Additionally, auto-tweets were generated when individuals signed the petition. Petitioning tweets were utilized the most, with 53% of the sample, or 988 tweets.

Information Release. Tweets that share information that is leaked, redacted, or otherwise considered confidential were coded as Information Release. It emerged that the tweets in this category focused on a single event: "Twoops: #CISPA co-sponsor deletes tweet highlighting his financial ties to pro-CISPA companies: http://snlg.ht/105gzIS." Again drawing on the replicability of the Internet, by tweeting and retweeting about the faux pas, users worked to keep the congressman's mistake alive, despite his attempt to delete it. There were 20 tweets of this nature in the sample, or 1.1%.

Information Source

To answer RQ3, links contained in #CISPA tweets were coded for source. The categories we created for source were informed by previous research (Bruns, Burgess, Crawford, & Shaw, 2012). There were 1,246 tweets with links. The source code indicates what type of outlet provided the information in the link. Intercoder reliability was very high at 98.3%. Six types of outlets emerged: Advocacy Groups, Citizen Media, Government, Political Blog, News Media, and Technology Media (see Appendix A for code descriptions). Figure 10.1 displays the frequency of sources that contributed to the information contained in the links.

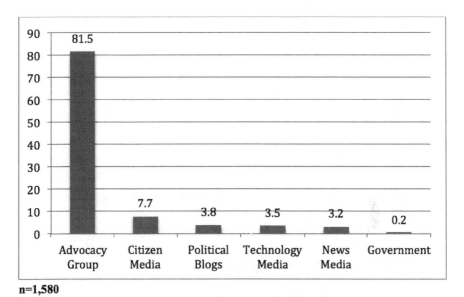

n=1,580

Figure 10.1. Sources of information in URLs.

Advocacy Groups were the source of information in 81.5% of the tweets that contained links. Comparatively, the government was effectively silenced in that it was the source of information in only .2% of the tweets.

DISCUSSION

This study examines the communicative and technical tactics used in Internet activism in **#CISPA**. We first identify the communicative tactics in tweets and then categorize tweets by identifying various technological activist actions, following Powell's (2011, 2012) prior categorization. These actions are then related to Van Laer and Van Aelst's (2010) typology of digital actions which entail varying levels of efforts and consequences.

RQ1 asks about salient communicative tactics based on the ICA framework. We identify that community and action are the two dominant tactics. However, in previous studies of Twitter content, most tweets were identified as informational (see Lovejoy & Saxton, 2012). We surmise that our unusual finding is due to the specific activism context of **#CISPA**. It is possible that users who participated in the Twitter discussions of CISPA were already informed about the legislation. Thus, they did not need and were not motivated to share information through tweets. Their focus was likely on building rapport (through support-related community tweets) and coordinating collective actions, particularly petition signing. Our results show that content produced by CISPA discussants was fairly engaging.

We identify that most TAAs were low threshold and Internet based, according to the topology proposed by Van Laer and Van Aelst (2010). Examples of such low-cost actions are petitions and Twitter mentions. The Twitter mention is a technology-afforded communication mode that features directedness (Xu & Feng, 2014). Previous Twitter studies have found the retweet (RT) function to hold particular significance (Bruns & Burgess, 2011; Meraz & Papacharissi, 2013). Thus, the fact that RTs were not significantly used in **#CISPA** sets it apart from previous hashtag research. Instead of retweeting, users employed the creative use of @mentions to hail—and in some cases, spam—politicians to effectively deliver a unified message. The use of @mentions may be considered a disruptive tactic if spamming is the desired outcome. However, it does appear as though the same "repetitious rhythm" found by Meraz and Papacharissi (2013) through RT was instead created by the continual use of @mentions in **#CISPA**.

Rather than rely on actions that are disruptive, the activism in **#CISPA** centered more on petition signing and @mentioning relevant politicians. Additionally, the lack of organizing tweets may indicate that **#CISPA** activists do not—in the early stages of activism—see value in organizing groups outside of signing petitions. This may be a result of the nature of the topic: to prevent CISPA from passing, activists need only to convince their local political representatives to vote against the bill.

In addition to their organizational function, hashtag streams make it possible for users to connect to a topic without necessarily connecting to other users (Bruns & Burgess, 2011). This technical function of Twitter makes it possible for nonreciprocal information sharing (Kwak et al., 2010), which helps to explain how hashtags may direct information flows. Previous Twitter studies have shown that news is co-constructed by participants, whether they are professional journalists, bloggers, or activists (Lotan et al., 2011). Yet by evaluating URLs in the **#CISPA** discussion network, it appears anti-CISPA activists dominated both the information flow and content. This was achieved in two ways: first, by completely shutting out the government's message, and second, by replacing it with one of their own. The lack of government-produced media in link sharing prevented the government's stance from appearing in the discussion network at all. Instead, users were primarily sharing links that were created by the activist groups themselves. This measure ensured that the position of the activist groups received more widespread attention in the discussion network.

LIMITATIONS AND FUTURE RESEARCH

Our sample represents a small portion of **#CISPA** tweets. Thus, the findings should be treated as indicative of potential trends. Additionally, there is no way to measure whether the actions performed on Twitter had a direct impact on CISPA's fate. Nevertheless, the use of TAAs in Twitter merits future research. Our findings indicate that sometimes a tweet is more than words; sometimes a tweet invokes technical aspects of the underlying network structure which offer leverage against institutions of power.

NOTES

1. The formula for determining this coefficient in terms of percentage of agreement is [2M/(N1+N2)], where M is the number of coding decisions on which two coders agree and N1 and N2 refer to the total number of coding decisions by the first and second coder, respectively (Wimmer & Dominick, 1997).
2. Many of these tweets were the result of "@fightfortheftr and @demandprogress tweeting some of over 1 Million signatures already collected by groups against **#CISPA**." Retrieved from https://twitter.com/CISPApetition

ACKNOWLEDGMENTS

This material is based upon work supported by the National Science Foundation under grant no. DGE-1069311.

Appendix A. Information Source Codes and Descriptions

Code	Information Source	Examples
Advocacy Group	Advocacy, Grassroots	ACLU, Anonymous, Electronic Frontier Foundation
Citizen Media	Individual	Personal Blogs, Individuals' Websites
Government	Government Offices or Agencies	Official Statements, Press Releases
News Media	Traditional Media Outlets	Newspaper, Radio, or Television
Political Blog	Political Bloggers	The Hill, Talking Points Memo
Technology Media	Technology Writers	TechCrunch, Wired

REFERENCES

Breindl, Y., & Briatte, F. (2013). Digital protest skills and online activism against copyright reform in France and the European Union. *Policy & Internet, 5*(1), 27–55.

Bruns, A., & Burgess, J. (2011). *The use of Twitter hashtags in the formation of ad hoc publics* [Author's version]. Paper presented at 6th European Consortium for Political Research General Conference, August 25–27, University of Iceland, Reykjavik. Retrieved from http://eprints.qut.edu.au/46515/1/The_Use_of_Twitter_Hashtags_in_the_Formation_of_Ad_Hoc_Publics_(final).pdf

Bruns, A., Burgess, J. E., Crawford, K., & Shaw, F. (2012). *#qldfloods and @QPSMedia: Crisis communication on Twitter in the 2011 South East Queensland floods.* Brisbane: ARC Centre of Excellence for Creative Industries and Innovation.

Carty, V. (2002). Technology and counter-hegemonic movements: The case of Nike Corporation. *Social Movement Studies, 1*(2), 129–146.

Chen, A. (2012, April 26). The non-geek's guide to CISPA, the cybersecurity bill the Internet is freaking out over. *Gawker.com.* Retrieved from http://gawker.com/5905081/the-non+geeks-guide-to-cispa-the-cybersecurity-bill-the-internet-is-freaking-out-over

Costanza-Chock, S. (2003). Mapping the repertoire of electronic contention. In A. Opel & D. Pompper (Eds.), *Representing resistance: Media, civil disobedience and the global justice movement* (pp. 173–191). London: Praeger.

Earl, J., & Kimport, K. (2011). *Digitally enabled social change: Activism in the Internet age.* Cambridge, MA: MIT Press.

Gillian, K. (2009). The UK anti-war movement online: Uses and limitations of Internet technologies for contemporary activism. *Information, Communication & Society, 12*(1), 25–43.

Gruzd, A., Wellman, B., & Takhteyev, Y. (2011). Imagining Twitter as an imagined community. *American Behavioral Scientist, 55*(10), 1294–1318.

Hawksey, M. (2013, February 15). *Twitter Archiving Google Spreadsheet TAGS v5.* Retrieved from http://mashe.hawksey.info/2013/02/twitter-archive-tagsv5/

Hermida, A., Lewis, S. C., & Zamith, R. (2014). Sourcing the Arab Spring: A case study of Andy Carvin's sources on Twitter during the Tunisian and Egyptian Revolutions. *Journal of Computer-Mediated Communication, 19*(3), 479–499.

Holsti, O. (1969). *Content analysis for the social sciences and humanities.* Reading, MA: Addison-Wesley.

Kwak, H., Lee, C., Park, H., & Moon, S. (2010). What is Twitter, a social network or a news media? *Proceedings of WWW'10.* Retrieved from http://dl.acm.org/citation.cfm?id=1772751

Lotan, G., Graeff, E., Ananny, M., Gaffney, D., Pearce, I., & boyd, d. (2011). The Arab Spring: The revolutions were tweeted: Information flows during the 2011 Tunisian and Egyptian revolutions. *International Journal of Communication, 5*, 1375–1405. Retrieved from http://ijoc.org/ojs/index.php/ijoc/article/view/1246/613

Lovejoy, K., & Saxton, G. D. (2012). Information, community, and action: How nonprofit organizations use social media. *Journal of Computer Mediated Communication, 17*(3), 337–353.

Marwick, A. E. (2011). I tweet honestly, I tweet passionately: Twitter users, context collapse, and the imagined audience. *New Media & Society, 13*(1), 114–133.

McAdam, D., Tarrow, C., & Tilly, C. (2001). *Dynamics of contention.* Cambridge, UK: Cambridge University Press.

McCullagh, D. (2013, March 13). Privacy backlash against CISPA cybersecurity bill gains trac-tion. *CNET.* Retrieved from http://news.cnet.com/8301-13578_3-57574196-38/privacy-back lash-against-cispa-cybersecurity-bill-gains-traction/

Meraz, S., & Papacharissi, Z. (2013, January). Networked gatekeeping and networked framing on #Egypt. *The International Journal of Press/Politics, 18*(2), 138–166.

Morozov, E. (2009). The brave new world of slacktivism. *Foreign Policy, 19*(05).

Papacharissi, Z., & de Fatima Oliveira, M. (2011). *The rhythms of news storytelling on Twitter: Coverage of the January 25th Egyptian uprising on Twitter.* Paper presented at the World Association for Public Opinion Research Conference, Amsterdam.

Powell, A. (2011). Emerging issues in Internet regulation: The unstable role of Wikileaks and cyber-vigilantism. In I. Brown (Ed.), *Research handbook on Internet governance.* Northampton, MA: Edward Elgar. Retrieved from http://ssrn.com/abstract=1932740

Powell, A. (2012). Assessing the influence of online activism on Internet policy-making: The case of SOPA/PIPA. *Social Science Research Network.* Retrieved from http://papers.ssrn.com/sol3/papers.cfm?abstract_id=2031561

Rolfe, B. (2005). Building an electronic repertoire of contention. *Social Movement Studies, 4*(1), 65–74.

Van Laer, J., & Van Aelst, P. (2010). Internet and social movement action repertoires: Opportunities and limitations. *Information, Communication & Society, 13*(8), 1146–1171.

Xu, W. W., & Feng, M. (2014). Talking to the broadcasters to broadcast, on Twitter—Twitter con-versations with journalists as a practice of networked gatekeeping. *Journal of Broadcasting & Electronic Media, 58*(3), 420–437.

Xu, W. W., Sang, Y., Blasiola, S., & Park, H. W. (2014, March). Predicting opinion leaders in Twit-ter activism networks: The case of the Wisconsin recall election. *American Behavioral Scientist, 58*(10), 1278–1293.

Art, Craft, and Pop Culture Hashtag Publics

CHAPTER ELEVEN

Realism against #Realness: Wu Tsang, #Realness, and *RuPaul's Drag Race*[1]

ANDY CAMPBELL

> In order to fall apart as complex beings, we need first to be able to live.
> —Wu Tsang, *CLASS* blog, November 18, 2011

> What you find, what you feel, what you know, to be real.
> —Cheryl Lynn, "Got to Be Real"

I

Out of the overexposed whiteness, usually indicating the end of a film reel, a space suddenly comes into focus: it is a dimly lit and mostly empty loft. Windows line the back wall, and the faint red glow of car taillights crisscrossing below can be seen through them. There are two sources of discernable light in the space, a few strips of warm yellow theatrical lights set low on the ground and pointed up to illuminate a white brick wall on the right and a softer, whiter light coming from off-screen left. This second light source illuminates a pile of bodies on the ground. Slowly—shakily—the camera moves towards these bodies.

Far from being strewn about casually, these people are placed, taking up partic-ular poses and positions. One person looks as if she is attempting to get up, another like she is napping, and yet another is on her hands and knees. Voices speak in rhythmic succession: "I believe that there's a big future out there. A lot of beautiful things. A lot of handsome men. A lot of luxury," and "I want my sex change," and "I want to get married in church, in white," and "I want my name to be a household product." The litany of their wants and desires conveys both gender-normative fantasies of marriage and class aspiration, as well as non-normative ones (getting a sex change).

As the camera moves closer, it becomes clear that the piled people are the ones speaking these desires. Yet there seems to be, only in fleeting moments, a time lag between the words spoken and the lips of the speaking. Perhaps the sound-sync is off, perhaps the film has been dubbed, perhaps the people are, in fact, not speaking at all.

Finally, of the five performers, the one on the far left speaks this text: "I want everybody to look at me and say 'There goes Octavia.'" This statement is notable for two reasons: the name of the performer is actually Wu Tsang, not Octavia, and he is also the artist and author of this particular film, *For How We Perceived a Life (Take 3)*.[2] Tsang's utterance confirms what savvy viewers will have already picked up, that the lines spoken by the performers are lifted directly from Jennie Livingston's 1990 documentary on the Harlem (drag) ball scene, *Paris Is Burning*. Tsang speaks the words of Octavia St. Laurent, one of the members of the ball community profiled by Livingston. Another member from the group breaks and begins delivering a pitch-perfect tirade, originally served by Crystal Labeija from the lesser-known 1968 documentary *The Queen*. Tsang's *For How…* is specific in archive because it focuses on the representation(s) of New York's Black and Latino trans communities in documentary film and practice.

"Full body quotation" is the name of the performance technique developed by Wu Tsang to simultaneously perform this archive and communicate the placed subjectivities of the performers. The technique involves a performer lip-syncing (via a hidden audio source) to their own voice re-speaking an appropriated text. This would not be so different from the kind of lip-syncing one might encounter at a drag show were it not for the fact that the performer is essentially lip-syncing themselves. The technique seems to involve a good-faith attempt on the part of the performer to capture the nuances of tone, accent, rhythm and breath of the person they're channeling, with the knowledge that the distance between the "original" audio and the "re-spoken" audio helps to define the break, the violent separation between subjectivities. When Tsang speaks the words of Octavia St. Laurent, Tsang approximates Laurent's full voice and stately rhythms but is not, in fact, Octavia St. Laurent. Through mimesis, Tsang "question[s] authenticity and intention of the speaker, and understand[s] content differently, out of its original context," evincing "a way to perform our ambivalences" (Wyma, 2012).

In this article, I wish to privilege texts that, in the words of Jack Halberstam, "do not make us better people or liberate us from the culture industry," but rather, "offer strange…logics of being and acting and knowing" (2011, pp. 20–21). In this spirit, I coordinate the reality show *RuPaul's Drag Race* (2009–), fully aware of its status as "kinda subversive, kinda hegemonic" (Sedgwick, 1993, p. 15), with the appended belief that Tsang's performance technique and film have the ability to limn such traits (subversive/hegemonic) as well as exceed them. Finally, I argue that Tsang's full body quotation, as an embodied performance technique, has the ability to counter the #realness often found in Twitter feeds and Facebook posts, supplanting it with an embodied—and markedly nondigital—queer realism. This is even in the face of the emergence of the "masspersonal" (Wu, Hofman, Mason, & Watts, 2011) "technocultural assemblage" of Black/Brown Twitter (Sharma, 2013; Clark, Chapter 15, and Cantey & Robinson, Chapter 16, this volume), which, vis-à-vis the performative process of its naming, demands identification with bodies that are both spectral and real, something which José Esteban Muñoz identified, in thinking through Tsang's work, as a "brown commons" (2012; see also Manjoo, 2010). In this way, Tsang's full body quotation is a way of countering the formal traits that give digital flow (to snipe from Williams, 1974) of data and metadata—whether they be hashtags, retweets, reblogs, links, etc.—cultural and economic value.

One hashtag, transposed from *Paris Is Burning*, seems to sum up these ambivalences: #realness. But to get to #realness, we might first want to understand the referent: realness, as it appeared in Livingston's documentary.

II

Paris Is Burning details the lives and performances of some participants in the Harlem drag ball scene—mostly gay Black and Latino men—in the late 1980s. Since its release, the film became "central to academic work…almost immediately" (Hilderbrand, 2014, p. 121). One of the key formal traits of Jennie Livingston's documentary is the insertion of title cards, often ball slang, which define particular segments; "REALNESS" is one such title card. Realness is an etymological riff on the real/reality; its character is that it is only *like* the real, approximate, near, but not quite.

Funnily enough, the subjects in *Paris Is Burning* don't agree on what the term exactly means. Both Junior Labeija (whose voice appears in voice-over but is never interviewed on camera) and Dorian Corey emphasize that realness's key feature is the ability to "blend" and to "pass the untrained eye, or even the trained eye." This is how the term is popularly (mis)understood, as an attempt to pass for a normatively gendered, straight person. There is a mountain of literature on passing and its complexities; needless to say, for Black/Latino gay men, passing is an historically loaded mode of being in the world, predicated on the traumatic conditions of daily life as a person of color. Realness, then, is actually rendered in sharp contrast to reality; in *Paris Is Burning*, realness is often parodic, a kind of deep gallows humor. As Dorian Corey later points out (in contradiction to his earlier statement):

> In real life you can't get a job as an executive, unless you have the educational background and the opportunity. Now the fact that you are not an executive is because of the social standing of life.... Black people have a hard time getting anywhere. And those that do...are usually straight. In a ballroom you can be anything you want. You're not really an executive, but you're looking like an executive. And therefore you're showing the straight world—I can be an executive. If I had the opportunity, I could be one. Because I can look like one. And that is, like, a fulfillment. (Livingston, 1990).

Throughout *Paris Is Burning* we see a variety of categories at the ball Livingston films: Schoolboy Realness, Town and Country, Executive Realness, High Fashion Evening Wear, Banjee Boy/Girl, Butch Queen First Time in Drags at a Ball, etc. In this sense, realness contains within it the critique of normative power structures, as realness is performed by those who are systematically denied access to the boardroom, the clubhouse, and such places that confer cultural or economic capital. When a contestant competing in the "Schoolboy Realness" category flops down with a choreographic flourish to read a book on the runway, the realness "tell" is not his preppy clothes or wire-rimmed glasses (both are lovely, immaculate renderings of a certain platonic ideal of studiousness), but rather that he begins to read his textbook...backwards.

Realness, at the moment when *Paris Is Burning* was filmed, comes with the simultaneous acknowledgement that despite the exactitude or virtuosity of a performance, systemic oppression is real and cannot be overcome by dress and/or gesture. I've always wondered why the trophies for the ball in *Paris Is Burning* were so large, some towering over the people competing for them: the world outside the ballroom doesn't give trophies to Brown/Black gay men and trans folks, and so the trophies, like the realness of the performances, are both an admission and refutation of intersectional oppressions.

For critics of *Paris Is Burning*, the concept of realness presents something of a sticking point. bell hooks's oft-quoted essay "Is Paris Burning?" describes *Paris Is Burning* as "a graphic documentary portrait of the way in which colonized black people (in this case black gay brothers, some of whom were drag queens) worship at the throne of whiteness" (1992, p. 149). hooks's analysis rests on the assumption (and for some of the subjects of *Paris Is Burning*, she is certainly on point) that they seek only to emulate whiteness instead of envisioning Black/Brown bodies in new constellations of power. True, the queens in *Paris Is Burning* do not offer a radical feminist critique of power—seeking to divest power at its foundations—and this may mean that their wants/dreams/desires are suspect to critics such as hooks who read them as wantonly collusive with hegemony. Yet the film also became an important touchstone for a generation of trans performers, who consistently reference the film in their art (Hilderbrand, 2014). One such performer is the mononymous "Supermodel of the World," RuPaul.

III

RuPaul's Drag Race (*RPDR*) is a reality television competition program developed by World of Wonder productions and broadcast by LogoTV (a subsidiary of Viacom). What began in 2009 as a rinky-dink show on a marginal cable channel (some of us still remember the jury-rigged studio set of the first season) is now an award-winning program that has seen viewership numbers topping 1 million (Gorman, 2013). The show's success is due, in no small part, to its meta-reality television format, riffing broadly off of tropes developed by other popular reality series such as *America's Next Top Model* (*ANTM*) and *Project Runway*. Indeed, RuPaul fulfills the roles of Tim Gunn (*Project Runway*) and Tyra Banks (*ANTM*) simultaneously by appearing in boy-drag as a mentor to his drag queen contestants during the first half and as a judge in girl-drag during the second half. The exaggeration of *RPDR* means that it "functions both as a parody of reality TV and as the most effective example of it on the airwaves" (Searle, 2013).

RPDR, World of Wonder, and LogoTV/Viacom build and maintain its audience through a sophisticated deployment of social media marketing which migrates content across platforms. Twitter accounts linked with the show, @ RuPaulsDragRace, @RuPaul, @MichelleVisage, @SantinoRice, @ WorldOfWonder, and @LogoTV, each have tens to hundreds of thousands of followers, and the new cross-promotional relationship with gay networking "hook-up" app Scruff dominated the sixth season. After the first few seasons, the show and its producers/distributors became savvy at using Twitter and Instagram, creating animated .gifs of the runway segments, introducing a "Dragulator" app for smartphones, suggesting playlists (featuring, who else? RuPaul!), initiating contests, and consistently juggling and cross-referencing the personalities of the star, producer, contestants, former contestants, and a loyal fan base. Twitter, and the use of hashtagging to connect fans to personalities, is an active part of forming and maintaining *RPDR*'s fan ecosystem. To stoke the flames of fandom and viewership, *RPDR*'s presence on Twitter (@RuPaulsDragRace) promises authentic and daily interaction with the show's host. Referring to fans as "children" (an overt reference to the ball communities represented in *Paris Is Burning*) and "squirrel friends," Ru acts as a house mother and playful best friend. It's a calculated persona to be sure, but one which has shown a steady increase in viewership and cultural capital for the show, its producers, and distributing cable channel/corporation—netting more awards and viewers with each successive season (TCA, 2014).

During the final episode of the sixth season, RuPaul broke from the usual Q&A format with the season's contestants and presented a montage-retrospective, not dissimilar to the "in memoriam" montages that are a staple of live awards shows such as the Academy Awards. This montage-retrospective operated as a kind of remedial lesson in drag history. In fact, the beginning slide of the montage-retrospective features the words "Drag History 101" written in lipstick on a dressing room mirror, one of the proprietary repeated rituals enacted by the departing queens of *RuPaul's Drag Race*. I would argue that this is a particularly potent moment to pay attention to because it highlights the historical moments that inform the show's present. In this "montage as history lesson," there are particular geographies of drag culture that are highlighted (a disproportionate amount of New York queens), credentials campily elided or fudged (Chi Chi LaRue is certainly a director, but perhaps more specifically a porn director), and strands of drag history never mentioned (no mention of drag kings or ciswomen in drag). What follows is a loose transcript of the live intro to the montage-retrospective as well as the voice-over (VO) text and the corresponding slides with the identified figures and their titles in brackets.

> From bearded ladies to mermaids, from club kids to pageant queens, this season the different styles of drag raised a lot of questions. So we put together a little herstory lesson. Learn it. And learn it well! [video clip starts, Rupaul VO:] What is drag? [Drag History 101] Drag is underground [Joey Arias; Chanteuse], and over the top [Tim Curry, *The Rocky Horror Picture Show*]; drag is political [Rupaul, March on Washington, 1993], and politically incorrect [Chi Chi LaRue, Director]; drag is camp [Lypsinka, Performer], and couture [Billy Beyond, Model]; drag is punk [Divine, Actress], and mainstream [Jimmy Fallon, Seth Rogen, and Zac Efron, *The Tonight Show*]; drag is a laugh-riot [Flip Wilson, Comedian], and it can start a revolution [Marsha P. Johnson, Stonewall Riots Activist]; drag is never having to say "sorry" [Lady Bunny, Filthy Bitch], because drag is all about being whoever the hell you wanna be [Holly Woodlawn, Warhol Superstar; Sylvester, Disco Diva; Kevin Aviance, Legendary Club Performer; John Cameron Mitchell, *Hedwig and the Angry Inch*; Neil Patrick Harris, *Hedwig and the Angry Inch*; Conchita Wurst, *Eurovision 2014* Winner; *Paris Is Burning*]. (World of Wonder, 2014).

Following the genre convention of the montage-retrospective, it is the last image that carries the most weight. In this instance the referent is not a person, per se, but *Paris Is Burning* as a whole. The film is not modified by a linguistic descriptor but is assumed to stand on its own. Following the montage-retrospective, RuPaul and the queens onstage grasp one another's hands and sing a "We Are the World" parody unity anthem. Poking fun at (and adhering to) the notion that a "shared" history brings people together, RuPaul's performance of drag history emblematically places RuPaul at the "political" center of drag history.

It is from the vantage point that I seek to understand the ways in which *RPDR* has been instrumental in disseminating lingo from *Paris Is Burning* through the Twitterverse. This is especially important to consider because of Logo's recent push to redefine itself as a contributing force to "mainstream culture" (Fine, 2012). Even RuPaul, in an interview with *The Guardian*, speaks of the flow of drag culture towards a "mainstream": "Because of our show, gay pop culture is pop culture in the mainstream. Everybody knows all of the terminology. It's really interesting for us to bring a lot of the old ideas and gay culture forward to pop culture mainstream" (Rogers, 2014).

But it is worth remembering that realness was a turn of phrase, amongst many others, lifted wholesale by RuPaul and the producers of *RPDR* (World of Wonder) from *Paris Is Burning*. This is not to say that the appropriation was not genuine or that it does not accurately reflect the ethos of the program (I think it does in many ways) but rather that *Paris Is Burning*, just by virtue of the frequency with which it is referenced subtly and outright, serves as an enduring companion text to *RPDR*. It is often assumed as a shared filmic text amongst its contestants and viewing audience. As one commentator remarked regarding *RPDR*'s cribbing of lingo from *Paris Is Burning*, "I don't feel like a sell-out, but part of a self-aware game" (Blankenship, 2012).

Yet even before *RPDR* disseminated **#realness** via Twitter, users were already using the hashtag. Importantly, the uses of **#realness** on Twitter are a marked departure from the kind of realness described in *Paris Is Burning*. For example, a tweet dated September 8, 2009, from @JorgeDaCurious reads: "I, too, am interested to see Tyra's real hair. So…who's watchin the Tyra Show today? **#realness #truth**." The episode of *The Tyra Show* in question witnessed "Tyra reveal[ing] her real hair; and talk[ing] with Perez Hilton" (TV Guide, n.d.). Here, as in most cases of **#realness**, the primary referent is an assumed, uniform truth or reality.

In this sense, RuPaul does have a politics as regards the use and dissemination of realness/**#realness**. Funnily enough, between RuPaul's Twitter accounts (@RuPaul and @RuPaulsDragRace), she has used **#realness** less than a dozen times, but realness hundreds of times. Indeed, RuPaul's 2015 album is titled (what else?) *Realness*. This, for me, speaks to a radical break between **#realness** and realness, and a pervasive argument waged against **#realness**. Reflecting their early history as "groupings," users who hashtag realness essentially join the myriad others who (mis)use **#realness** as an analogue for the real, not always as an embedded commentary upon the structural inequities of social life (Coates, 2013).

Realness is a near-constant exclamation on *RPDR*. As the queens compete in challenges ranging from the benign to the silly, they are ultimately judged on their performances in the challenge and a subsequent runway look. As the contestants walk the runway, judges and fellow contestants (via voice-over) comment upon their performances. This segment hews most closely to walking a ball in

Paris Is Burning, and unsurprisingly it is here that realness is often brought up. As *RPDR* has moved from season to season, the realness of the imagined "categories" has gotten ever-more baroque. In describing their runway looks, the contestants and judges of the first episode of the fifth season, "Rupaullywood," invented the following categories: Real Housewife Realness (Camille Grammer, a character on Bravo's *Real Housewives of Beverly Hills*, appears later in the episode); Hellen Keller Drowning Realness; Daytime Realness; Dark and Twisted Alyssa Edwards Realness. Instead of conforming to or riffing on a pre-established ball category, they create the categories they are already winning.

The oscillation between embedded understandings of realness and **#realness** is a narrative structure carefully established by RuPaul and the *RPDR* producers from the very first episode. The show's introductory montage/monologue finds RuPaul giving an accounting of his life ("I, RuPaul, was born a poor black child in the Brewster Housing Projects of San Diego. But, baby, look at me now!"), lets viewers know what they can expect from contestants ("and just when you think these queens have gone as far as they can go, they push it one step further"), and makes the first mention of the real ("Are you ready for the ride of your life? *Rupaul's Drag Race* is about to get real in 3…2…1…") (World of Wonder, 2009). The real is on notice as soon as the first contestant (Shannel) enters the workroom wearing assless zebra-print tights, describing her drag persona as "more on the realistic side."

IV

i could go on for days about the nuances, but basically this work is about exploitation. not like in a judgmental way, but like a rubiks-cube way, like these are all the sides to the problem—there is no easy solution. and there are SO MANY ways it can go wrong…it's a really tight spot.

—Wu Tsang, *CLASS* blog, November 18, 2011

I began this article with a description of *For How…* because it makes use of material from *Paris Is Burning* and actively compounds its meanings. Tsang's full body quotation takes equally from historical modes of Trans/drag performance (lip-syncing) and newer modes of sharing information (the hashtag, the retweet/reblog) to create an embodied modality which defies easy commodification within either realm. Full body quotation might be too "arty" for a typical drag show, or too messy to be encapsulated in 140 characters or less. Its value, then, is in being between these placements—of performing its own ambivalence. I want to suggest that the value of Tsang's performance is tied to its resistance, perhaps even failure, in the realm of hashtag publics—a realm that *RPDR* succeeds in.

The above text concerning the "really tight spot" of performance was published on Wu Tsang's production blog the day before he and a small group of collaborators premiered the performance *Full Body Quotation* at the New Museum (New York City), as part of RoseLee Goldberg's live performance festival, Performa 11. Described as "a choral performance" and a "living sculpture"—both terms that situate Tsang's practice within particularized arenas of theater and performance art—*Full Body Quotation* predated *For How...* but shares much in common with it. Tsang collaborated with the same performers and used the same sampled dialog for both works. On his blog, Tsang writes openly of his ambivalence towards performing in the Sky Room of the New Museum—a space described in the New Museum's marketing materials as its "premier space." Tsang, for his part, marks the space as "not necessarily a safe space for all the communities referenced in this work." Tsang's online post was accompanied by a curious graphic—a film still of Octavia St. Laurent, one of the focal characters from *Paris Is Burning*, with the overlying text: "99% OF GAY WHITE MALES MISUSE 99% OF THE DIALOG FROM *PARIS IS BURNING*. STOP SAYING *REALNESS*. #OCCUPYBALLROOM." While the graphic is not of Tsang's creating (it was found by the artist on howtobeafuckinglady.tumblr.com), it clearly demarcates one of the "tight spots" of Tsang's particular archive and makes direct reference to the then-ubiquitous protest language of the Occupy movement and its legitimacy within hashtag publics.[3] #OCCUPYBALLROOM seems to have been a call, but without response; an intended meme that never went viral. Whoever made the graphic (its authorship is unclear to me) is attempting to knit together the concerns of the #occupy movement with a perceived large-scale appropriation of Harlem ballroom culture lingo.

Much of the conversation surrounding hashtags, indeed, much of the scholarship in this volume, puts them in relation to their virality—the ability to disseminate broadly with great velocity; #occupy and #winning are examples. #OccupyBallroom, however, is a sterling example of the other side of this technical affordance: the failed hashtag; at time of writing, it has barely been tweeted into the double digits. Twitter assigns value (cultural/economic) to tweets and users who meet the algorithmic requirements for velocity, shareability, and geographic location (Laird, 2013; Twitter, n.d.). But even when Twitter discusses its algorithms, it's difficult to know how to assess hashtags because, as Virginia Heffernan writing for the *New York Times* in 2009 astutely noted, hashtags "are almost never transparent or ideologically neutral" (p. 13). On most of these fronts, #OccupyBallroom underperforms, and thus, we might not wonder why it is therefore banished into Internet obscurity. Comparatively, #realness was tweeted hundreds of times in the month of Tsang's post. But the curious thing about the graphic posted by Tsang is that even though the hashtag didn't travel, the graphic itself did: the still of Octavia St. Laurent with appended text was retumbled over 400 times.

What Tsang's blog, performance, and film suggest is that we must also think about the informational logics that undergird our conception of what's worth paying attention to and be wary of overly utopic envisioning of digital lives as unrelated to the corporeal bodies we inhabit daily. Tsang has constructed a thoughtful counterpoint to the kinds of consuming publics drag lingo hashtags (such as **#realness**) seem to be symptomatic of and enact. That drag culture and lingo has been thoroughly commodified and resold as a (too) neat counterpublic/ radical package is perhaps not news, but its imbrication within new modes of dissemination (Twitter, Instagram, Facebook, Tumblr) serves as continuing evidence that the uneven distribution and accretion of cultural capital is predicated on trans/drag lives.

V

Finally, I want to suggest that instead of **#realness**, a term that has become so ubiquitous as to be only tenuously connected with its source, "realism" might be a more apt term for Tsang's performance and video. No hashtag.

Realism captures the politics/representation knot of realness as well. Because Tsang is a self-identified artist, his work also enters the discourses surrounding the history of art and criticism. And within these discourses, the lineaments of realism as a strategy (as well as a particularized movement/style in 19th-century Europe) are under near-constant negotiation. Writing of realism in painting, the British critic Lawrence Alloway wrote, "the depicted objects have to be continuous with the symbolic system of ideology that they signify" (2006, p. 228). And this tactic, as art historian Linda Nochlin points out, has been historically "reserved for the representation of lower orders of humanity" (1973a, p. 58). This would have included subjugated and lower-class workers in the context of 19th-century France—and in Tsang's performance it extends to queer people of color. That Tsang uses lip-syncing, an historically recognized arena of trans/drag performance, marks his performance technique as "continuous with the symbolic system of ideology that [it] signifies," and thus is an extension of transcultural forms and concerns regarding representation.

"Yet realism," for Linda Nochlin, "involves both more and less than visual veracity" (Nochlin, 1973b, p. 98); "more than" because rendering a thing without an extra gloss is a nearly an impossible task, "less than" because it is not the things themselves being represented but "the authenticity of the contiguous relationships existing amongst concrete figures" (Nochlin, 1973b, p. 98). Full body quotation conforms to Nochlin's description as the performers approximate—à la realness—the voices they channel, without fully slipping subjectivities. The performers are still themselves; they do not become Octavia St. Laurent or Venus Xtravaganza. Their conceit and surroundings are markedly self-reflexive and theatrical.

This is what Raymond Williams referred to as the "shifting double sense" of the real (1976, p. 258), asserted most clearly in Tsang's performance syntax, which through close, limited, and sustained collaboration devalues the kinds of criteria that have, as of yet, defined the cultural importance of a hashtag—namely, the statistical count of people who use, retweet, riff, and negate a particular hashtag and the velocity of its distribution. Tsang's full body quotation, as technique and as particularized performances, contains and requires a slower/longer time scale. The formal piling of bodies in performance, of quotations from the history of trans cinematic representation, of geographical classed space, and of performance modalities (live vs. filmed) enable a viewer to consider a marked alternative to the hashtagged words of drag queens and their lingo as exemplified by *RPDR*.

Furthermore, Tsang builds coalitional possibilities between trans and drag communities—an anxious divide that is now a repeated storyline spectacle on *RuPaul's Drag Race* as contestants come out as trans (as though they weren't already included under the broader brim of that umbrella as drag queens). Tsang, in short, presents the realism of **#realness**.

Towards the end of *For How...*the five performers, who are of varied colors, body types, and genders, line up against the white brick wall and sync: "this movie is about...supposedly about...supposedly about." This collective intonation is a trickster-like polysemic event—which film? Tsang's? Livingston's? And if things aren't what they seem—if the film is not what it is supposedly about—what might it be about? No time to think: the camera turns towards the darkened back end of the room, which disappears into velvety darkness. The film reel runs out, a brief flash of white light, and Tsang's film loops back to the beginning.

NOTES

1. Correspondence concerning this article should be sent to Andy Campbell, 318D Hyde Park Blvd., Houston, TX, 77006. andycampy@gmail.com
2. I attribute the work to Tsang, but this is only shorthand, as the artist is emphatic about his work often being the product of collaborative and collective modes of making. I want to thank Tsang for his lucid commentary and insight about this work, as well as for providing me access to *For How to Perceive a Life (Take Three)*.
3. I have yet to track this graphic's authorship; the earliest example of it I've found was on the Tumblr of sabao-spray.

REFERENCES

Alloway, L. (2006). Realism as a problem. In R. Kalina (Ed.), *Imagining the present: Context, content, and the role of the critic* (pp. 227–230). Oxford, UK: Routledge. (Original work published in 1974)

Blankenship, M. (2012, April 23). Mopping drag slang. *Out*. Retrieved from http://www.out.com/entertainment/2012/04/23/drag-queen-rupaul-race-language-slang

Coates, T. (2013, May 7). The uncomfortable class connotations of *RuPaul's Drag Race*'s cultural appropriation. *Flavorwire*. Retrieved from http://flavorwire.com/390058/the-uncomfortable-class-connotations-of-rupauls-drag-races-cultural-appropriation

Fine, R. (2012, March 2). Logo network bails on gay-centric TV programming. *San Diego LGBT Weekly*. Retrieved from http://lgbtweekly.com/2012/03/02/logo-network-bails-on-gay-centric-tv-programming/

Gorman, B. (2013, January 29). *Season debut of "RuPaul's Drag Race" scores as the highest-rated premiere in Logo's network history* [Press release]. Retrieved from http://tvbythenumbers.zap2it.com/2013/01/29/season-debut-of-rupauls-drag-race-scores-as-the-highest-rated-premiere-in-logos-network-history/167327/

Halberstam, J. (2011). *The queer art of failure*. Chapel Hill NC: Duke University Press.

Heffernan, V. (2009, August 7). Hashing things out. *The New York Times*. Retrieved from http://www.nytimes.com/2009/08/09/magazine/09FOB-Medium-t.html?_r=0

Hilderbrand, L. (2014). *Paris is burning: A queer film classic*. Vancouver: Arsenal Pulp Press.

hooks, b. (1992). *Black looks: Race and representation*. Boston: South End Press.

Laird, S. (2013, February 4). *Twitter itself will soon decide the value of your tweets*. Retrieved from http://mashable.com/2013/02/14/twitter-judge-value-tweets

Livingston, J. (Producer/Director). (1990). *Paris is burning* [Motion picture]. USA: Miramax Films.

Manjoo, F. (2010, August 10). *How Black people use Twitter*. Retrieved from http://www.slate.com/articles/technology/technology/2010/08/how_black_people_use_twitter.html

Muñoz, J. E. (2012, October). *The brown commons: The sense of wildness*. Paper presented at Harry Ransom Humanities Research Center, University of Texas. Austin, TX.

Nochlin, L. (1973a, September–October). The realist criminal and the abstract law. *Art in America*, 54–61.

Nochlin L. (1973b, November–December). The realist criminal and the abstract law II. *Art in America*, 96–103.

Rogers, K. (2014, February 24). Rupaul: *Drag Rrace* "has exactly the effect we thought it might have." *The Guardian*. Retrieved from http://www.theguardian.com/tv-and-radio/2014/feb/24/rupaul-drag-race-lgbt-impact-pop-culture-tv

Searle, J. (2013, March 1). Just between us girls: Why *RuPaul's Drag Race* is the king (and queen) of reality television. *The Oxonian Review*. Retrieved August 11, 2014 from http://www.oxonianreview.org/wp/just-between-us-girls-why-rupauls-drag-race-is-the-king-and-queen-of-reality-television

Sedgwick, E. K. (1993). Queer performativity: Henry James's *The art of the novel*. *GLQ*, (1)1, 1–16.

Sharma, S. (2013). Black Twitter? Racial hashtags, networks and contagion. *New Formations, 78*(2), 46–64.

Simon, F. (Director). (1968). *The queen*. United States: Evergreen Film.

TCA. (2014). The Television Critics Association announces 2014 TCA awards winners [Press release]. Retrieved from http://tvcritics.org/the-television-critics-association-announces-2014-tca-awards-winners/

Tsang, W. (2011, November 18). Full body quotation at the new museum [Web log post]. *CLASS.* Retrieved from http://www.wildnessmovie.com/class-blog/2011/11/18/full-body-quotation-at-the-new-museum.html

TV Guide. (n.d.). Episode guide: *The Tyra Show.* Retrieved from http://www.tvguide.com/tvshows/the-tyra-show-2009/episode-2-season-5/premiere-show/195018

Twitter. (n.d.). FAQs about trends on Twitter. *Twitter Support.* Retrieved from https://support.twitter.com/articles/101125-faqs-about-trends-on-twitter

Twitter. (2010, December 8). To trend or not to trend…. [Web log post]. *Twitter blog.* Retrieved from https://blog.twitter.com/2010/trend-or-not-trend

Williams, R. (1974). *Television: Technology and cultural form.* London: Fontana.

Williams, R. (1976). *Keywords: A vocabulary of culture and society.* New York: Oxford University Press.

World of Wonder (Producer). (2014, May 19). The finale. In *RuPaul's Drag Race.* New York: Logo TV.

World of Wonder (Producer). (2009, February 2). Drag on a dime. In *RuPaul's Drag Race.* New York: Logo TV.

Wu, S., Hofman, J. M., Mason, W. A., & Watts, D. A. (2011). Who says what to whom on Twitter. In *Proceedings of the 20th International World Wide Web Conference (WWW 2011),* (pp. 705–715). New York: ACM. Retrieved from http://www.cs.cornell.edu/~sw475/

Wyma, C. (2012, March 2). "I dislike the word visibility": Wu Tsang on sexuality, creativity, and conquering New York's museums. *Artinfo.* Retrieved from http://www.blouinartinfo.com/news/story/761447/i-dislike-the-word-visibility-wu-tsang-on-sexuality-creativity-and-conquering-new-yorks-museums

CHAPTER TWELVE

Living the #Quilt Life: Talking about Quiltmaking on Tumblr

AMANDA GRACE SIKARSKIE

Thetwotwoone is a self-described "mom trying to make her living quilting and crocheting from home" (Thetwotwoone, 2014a). She is also a blogger on Tumblr who has employed many quilt-related hashtags, including the popular **#quilt life**. While this mother and her **#quilt life** might seem like strange bedfellows alongside users employing tags such as **#trans***, **#goth**, **#metal**, and the like, quilt- and craft-related hashtags, like the subaltern ones, serve similar functions of community building, self-identification, and in-group communication.

This essay will explore the use of hashtags in communicating about sewing and crafting, especially quiltmaking, on Tumblr. Crafters use different sorts of hashtags for different purposes. For example, thematic hashtags such as **#quilts**, **#sewing**, **#embroidery**, **#scrapbooking**, and the like can be used to alert others to the thematic content of the post. Commentary hashtags such as **#quilty pleasures** and **#quilt life**, in contrast, are often humorous or ironic and alert others to the commentary rather than the thematic nature of the post. Other hashtags, tags that I call "quilt-specific," such as **#sunbonnet sue** and **#grandmothers flower garden**, are used to generate community and conversations, as well as to create a sense of a quiltmaker community on Tumblr. Besides containing a general study of the use of hashtags by quiltmakers, this essay also explores the intersection of the quilt and craft community, via the use of shared hashtags,[1] with other sorts of communities on Tumblr, particularly fandoms. What can the use of hashtags by fans of various pop cultural phenomena tell us about the rhetoric of tagging on Tumblr? Finally, this essay will present a qualitative analysis of a selection of posts tagged with various hashtags. What does it mean, exactly, to live the **#quilt life** online?

It bears noting that studies on quiltmakers' communication practices online are hardly new. Judy Heim's *The Needlecrafter's Computer Companion: How to Use Your Computer for Sewing, Quilting, and Other Needlecrafts* (1995) and Heim and Hansen's *The Quilter's Computer Companion* (1998) were among the first works to explore this field.[2] I had initially intended this essay to be a study of the use of hashtags among quiltmakers on Twitter, analyzing hashtags such as **#quilter-problems,** a popular tag among quiltmakers on that site. And indeed, Twitter has become a quiltmaker's paradise; as Kristina Tabor wrote in a 2011 post on her blog, *The Nerdy Sewist,* "Twitter is a great resource if you're an avid crafter. I've learned so much about sewing, design, fabrics, even thimbles, machines, quilt squares, and much much more via Twitter. One of the best things about what I'll call 'Twitter for Crafters' are hashtags" (Tabor, 2011). I also considered focusing this essay on Instagram, where the hashtag **#makeaquiltmakeafriend** is de rigueur. In the past several months, however, I have discovered Tumblr—a dangerous hobby for an early-career academic already strapped for time—and the joys of the endless "reblogging" of image- and text-based posts pertaining to one's interests. Here, users (many but certainly not all of whom seem to be teenagers) apply hashtags to allow other users to search and find images and text-based posts, as well as to provide commentary on the posts.

With so many younger users on Tumblr, starting a blog can be daunting for older users. Tumblr's relatively high learning curve when compared to Twitter can make older novice users question their place on the site or their ability to start a blog altogether. In the case of Thetwotwoone, it is clear that the ambition to start a blog on Tumblr was tempered with feelings of inadequacy.

> I woke up the other morning and said to myself, "I should start a blog."
>
> I got up, let the dog out, and sat down to rouse myself with a cup of coffee and I realized—I don't know the first thing about writing a blog.
>
> I'm a stay-at-home mom of 4, running a small crafting business out of my dining room, barely squeaking by with the housekeeping and surviving solely on caffeinated soda, coffee and peanut butter sandwiches since my youngest was born and a good night's sleep moved out to make room, when would I ever have the time?
>
> And even if I did have time, what the hell would I say? **#quilting #crafts #life** (Thetwotwoone, 2014a)

As evidenced by this post, Tumblr can be quite isolating for new users. Hashtags, however—especially quilt-specific hashtags, which I will discuss in the following section—serve as a primary means by which quiltmaker-bloggers on Tumblr connect with other bloggers of like interests and begin to build a sense of community within the site.

#QUILT: HASHTAGS AND THE QUILT COMMUNITY

Of course, the most basic tags for quilt-related posts are **#quilt, #quilting, #quilter,** etc. These tags are highly useful in identifying the subject matter of a post as being quilt related. A quick search of Tumblr for **#quilt** predictably yields hundreds of image- and text-based posts about quilts. These tags are much less helpful, though, in marking out a **#quilt** community. This is because while some users devote the bulk of their blogs to quiltmaking or other crafts, many users with more general blogs, or indeed blogs about another topic altogether, might still post something relating to quilts and tag it as such. For example, on a general, slice-of-life blog, one user posted, "when im [sic] old im gonna turn all my band shirts into quilts and use them as bedding for my children **#band #band tee #quilt #5os #5 seconds of summer #paramore #sleeping with**" (5sos-stole-my-underwear, 2014).

For the user searching for a particular kind of quilt amongst the innumerable **#quilt** posts, however, many of the most helpful quilt-related tags describe the pattern, fabric, technique, or other visual or physical aspects of the quilt's design or construction. Popular tags in this vein include **#batik, #crazy quilts, #patchwork, #applique, #whitework, #redwork, #grandmother's flower garden, #sunbonnet sue, #log cabin, #lone star, #churn dash, #new york beauty, #whigs defeat,** etc. Indeed, in searching Tumblr over the period of some hours, I was hard pressed to find a known, named quilt pattern (as described in Barbara Brackman's seminal 1993 book *Encyclopedia of Pieced Quilt Patterns*) that had *not* been used as a hashtag on Tumblr.

Importantly, it is these tags, ones which make use of the quilt world's vocabulary outside of Tumblr, that most readily identify and demarcate the quilt community on the site. Sociologist Robert Redfield in *The Little Community: Viewpoints for the Study of a Human Whole* (1955) defined "the little community" as a small, distinctive, homogenous, and self-sufficient organization of people (p. 4). One sees this community aspect perhaps most strongly in posts in which the blogger is looking for information or assistance that other quiltmakers could provide. For example:

> I'm in need of some help. My mom has been searching for a long time for a particular Sunbonnet Sue pattern. It was in a book published by Martingale & Company, 1993. The name of the book was "Sunbonnet Sue All Through the Year." The problem is she has the patterns for 2 full-size quilts that came with the book, but someone had made another version of the quilt. It's the December square for Christmas and it has Sunbonnet Sue standing outside of a store window looking in at the display. I tried contacting the publisher and even searching the name of the quilter "LaVena Hallin." No luck. Does anyone out there know of this pattern or have it? I would GLADLY pay for the pattern. It would make my mom so happy. Thanks Tumblr Friends! **#Sunbonnet Sue #Quilt #Christmas.** (Nadine64, 2014)

Tags such as **#sunbonnet sue**, which have instant meaning for quiltmakers, truly do help to define a "little community" of quilt enthusiasts on Tumblr.

Hashtags are applied by quiltmakers not only to build community, however. Hashtags are also frequently used for the purposes of marketing quilt patterns, books, magazines, and the like. Jaybird Quilts, the label of California-based quilt pattern designer Julie Herman, actually prints the suggested hashtag for each of their quilt designs on the back of each pattern, encouraging quiltmakers who execute the patterns to post photos, using the appropriate hashtag, on Instagram, Twitter, Facebook, Pinterest, Flickr, or Tumblr (Herman, 2013). Jaybird hashtags as of August 2013 included:

Pattern Name	Hashtag
Opposites Attract	#OppositesAttractQuilt
Dance Floor	#DanceFloorQuilt
Dot Party	#DotPartyQuilt
Unwind	#UnwindQuilt
Firecracker	#FirecrackerQuilt
Hugs & Kisses	#HugsandKissesQuilt
Fast Forward	#FastForwardQuilt
Carnival	#CarnivalQuilt
Plaid Parade	#PlaidParadeQuilt
Off the Rail	#OffTheRailQuilt
Chopsticks	#ChopsticksQuilt
Taffy	#TaffyQuilt
Jawbreaker Pillow	#JawbreakerPillow
Biscuit	#BiscuitPincusion
Come What May	#ComeWhatMayQuilt
Radio Way	#RadioWayQuilt
Varsity	#VarsityQuilt
Yummy	#YummyQuilt
Ditto	#DittoQuilt
Ballerina	#BallerinaQuilt
Three in a Box	#ThreeInABoxQuilt
Teacups	#TeacupsQuilt
Northern Lights	#NorthernLightsQuilt
Wonton	#WontonQuilt
Candy Dish Pillows	#CandyDishPillow
Tasty Table Runner	#TastyQuilt

Traffic	#TrafficQuilt
Lotus	#LotusQuilt
Science Fair	#ScienceFairQuilt
Toes in the Sand	#ToesInTheSandQAL
Snack Time	#SnackTimeQuilt
Tiny Dancer	#TinyDancerQuilt
Giggles Baby Quilt	#GigglesQuilt
Seaside Table Runner	#SeasideQuilt
Rock Candy Table Topper	#RockCandyQuilt
Hex N More	#HexNMore
Sidekick	#SidekickRuler
Skip the Borders	#SkipTheBorders

Almost all of Jaybird Quilts' tags contain the word "quilt" within the hashtag, making the tags useful more or less only within the quilt community.

Quilt world hashtags are not foolproof in defining the blog posts of the quilt community, however. While **#sunbonnet sue** and many other quilt-related tags generally do return a high proportion of quilt-related posts, some quilt-related tags do not. A search of **#new york beauty**, for example, tends to return images not of New York Beauty quilts but of beautiful women in New York City. Another example is **#longarm**, which is used by quiltmakers to describe a certain type of machine (rather than hand) quilting done by a large (around 12-feet-long) semi-industrial sewing machine called a "longarm." These longarm sewing machines are available for use at many quilt shops and provide huge time savings over traditional machine quilting. The hashtag **#longarm** is also (and much more frequently) used by fans of the *Transformers* universe of action figures, television and film productions, comics and the like, because Longarm is one of its main characters. This dual use of the same tag by two presumably very different groups of people for two very different meanings can occasionally make searching by hashtags frustrating for users. I was indeed momentarily perplexed when I first looked for longarm-quilted textiles only to find dozens of images of a blue and silver robot.

#QUILT FANDOM

Besides these general and quilt-specific hashtags, on Tumblr as on Twitter, many quilt-related hashtags tend to be a bit cute or quaint, making use of puns or portmanteaus. One such tag is **#quilty pleasures**, used on Tumblr to describe a variety of quilt-related posts, including images of a very traditional red and green calico Christmas tree skirt, as well as the work of contemporary art quilter Luke Haynes (who maintains a significant presence in various social media channels). Confusingly, as with the case of **#longarm**, the tag **#quilty pleasures** is also used by the fandom of American singer and actor Zac Efron, and so a search for that tag yields many images and posts about him as well.[3]

It should be noted that quilt- and craft-related blogs, while quite abundant on Tumblr, are far outnumbered by Tumblr's golden quadrangle of food, humorous animals, sex, and fandom blogs. Hashtags adopted by fan communities are often particularly emblematic of tagging practices on Tumblr as a whole. Many of the fan hashtags on Tumblr are quite straightforward (e.g., **#Keith Richards**), and these, like **#quilt** for quilts, are the most useful for the user who is utilizing the tags for searching purposes. Other tags combine identification of the subject with description (e.g., **#Keithrichardsisgod** or **#keith is so out of it and mick has charm on full blast**), straddling a middle ground between utility and expression.

Given that one of Tumblr's primary uses is as a vehicle for fans of various works of film and television, literature, and music to blog and reblog content of interest to their fandom, **#quilt fandom** has emerged on Tumblr as a tag describing where these two worlds—quiltmaking and fandom—intersect. One user posted images of several quilt blocks decorated with embroidered outlines of the Doctor, the title character of the British science fiction series *Doctor Who*, done in a redwork style, but with white, black, blue, or yellow thread. The caption reads, "This is one of the best shows ever and now they are in quilt form for you!!! **#quilt #docter** [*sic*] **who quilts #docter** [*sic*] **who #quilt fandom**" (Marrow-j-houndzower, 2013). Depicted in the post is a quilt block featuring an appliquéd TARDIS (the space- and time-travel vehicle from *Doctor Who*) made out of felt. Another quiltmaker put out an "open call for someone who is perhaps overly obsessive with regards to both fandom and quilting because I need advice and someone who won't judge. **#quilting #I just don't know how I'm going to quilt this fucker #too many shades #and I can't think of a cotton that will look okay with them all #and I can't free motion on my shitty old machine #and I don't know any long-armers who arent fudgy old ladies**" (Pixel Quilts, 2014).

LIVING THE #QUILT LIFE

Turning away from the somewhat narrow usage of **#quilt fandom**, another way of using hashtags on Tumblr is to combine one or more content-related tags, such as **#quilt**, with the tag **#life** (or the variant, **#real life**). The added tag of **#life** cues readers to understand the post as a slice of the blogger's daily life. Thetwotwoone, the "mom trying to make her living" described in the introduction to this essay, posted:

I am always working on a quilt. If I'm not on my machine with it, I am designing a new one, researching fabric, attempting to learn a new technique.. it never ends.

My house is never clean anymore, and my family has had to forfeit the dining room table, but I am happy!...

My next post will be a walkthrough of making that very first quilt, wooo!

#quilting #handmade #diy & crafts #crafts # life (Thetwotwoone, 2014b)

This way of tagging is used by advanced quiltmakers, as seen in the post above, as well as by much more novice quiltmakers. For example: "Attempting to make a quilt! If the blocks come out anywhere close to square I'll count it as a win XD **#real life #quilts #quilting #look what I made #arts and crafts #sewing**" (Arora-kayd, 2014). That same novice quiltmaker later posted an update on their first quilt project: "Progress! **#real life #look what I made #quilting #the blocks are so uneven #*sobs***" (Arora-kayd, 2014).

The tag **#look what I made** in Arora-kayd's post shows that the user is clearly proud of their work, despite the quilt blocks being uneven. **#Look what I made**, incidentally, is another quite interesting tag because it is used by a variety of makers working with *very* different materials and techniques. Makers of more traditional material culture objects—quiltmakers, sewists,[4] woodcarvers, painters, and the like—employ the tag, as do users who have created a born-digital work of which they are proud. A search of **#look what I made** on June 17, 2014, turned up on the first page of the search results such diverse works as a gif of a scene in the television series *Bob's Burgers*, a photograph of a cheerleader pyramid, and a Lay's Potato Chips "Do Us a Flavor" contest entry titled "Dean Winchester" with "ingredients" including daddy issues, self-hate, and gay thoughts.

Returning to quilts, **#life** posts can also signal a life lesson, a life "pro-tip," or be used as a call for sympathy in the face of difficulty that the blogger is having.

pretty sure I'm going to end up destroying all of the nerve endings in my fingertips soon ironing and finger pressing

it does things to you

#it hURTS BUT IT'S NECESSARY #QUILT LIFE." (Libby-on-the-label, n.d.)

The post above combines a bit of all of these attributes, and by appealing for sympathy for a common plight that most quiltmakers share—namely, destroyed nerve endings in one's fingers—it creates a sense of camaraderie and community. The other tag used in this post, "**#it hURTS BUT IT'S NECESSARY**," along with the line in the previous post, "My house is never clean anymore, and my family has had to forfeit the dining room table," suggests that for this and other quiltmaker-bloggers, living the **#quilt life** is worth a few sacrifices in terms of personal comfort.

CONCLUSIONS

While this essay has focused on hashtags as used by quiltmakers as a means of connecting and communicating with other like-minded crafters on social media sites such as Tumblr, it is important to note that the simple visual geometry of the hashtag recently has been used also as artistic inspiration for quiltmakers, as evidenced by the "42 Hashtags" quilt pattern by Tanya Finken that appeared in the spring 2014 *Scrap Quilts* issue of the popular magazine *Fons & Porter's Love of Quilting*.

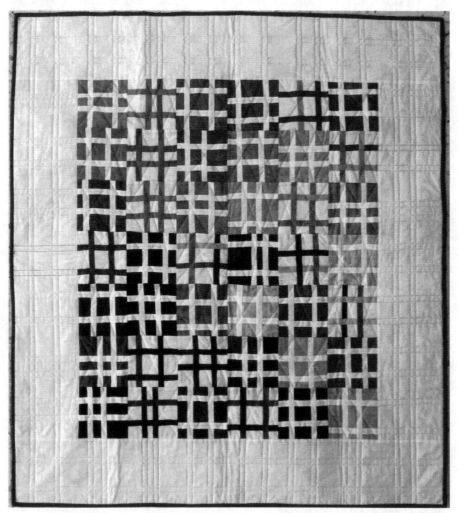

Figure 12.1. Lois Warwick, 42 Hashtags quilt, as seen on The Quilt Works, Inc., based on a pattern from *Fons & Porter's Scrap Quilts* magazine, spring 2014.

The pattern became popular quickly and has already been block of the month for the Seattle Modern Quilt Guild since its publication in spring 2014. The caption accompanying the hashtag quilt in *Fons & Porter's* reads, "This little quilt has an improv feeling. The designer's challenge was to use just one package of charm squares [pre-cut 5- x 5-inch quilt squares] to make her quilt. This throw-sized quilt features 42 multicolored hashtags (pound signs) on a white background" (Finken, 2014). The improvisational effect of Finken's scrappy design mirrors the spontaneous and free discursive practices of users on sites such as Tumblr and Twitter.

Hashtags can and do serve as means by which subaltern groups communicate and build identity. Hashtags, for example, are used to coalesce subaltern groups or as vehicles for organizing and mobilizing social protest. The hashtag, however, has also been adopted as both a marker and mode of discourse by more mainstream or traditional publics, such as quiltmakers. The **#quilt life** does, I think, provide a very interesting counterpoint to some of the other essays in the collection. For those living the **#quilt life**, hashtags facilitate communication about the banal, the domestic, and the cuddly.

NOTES

1. These hashtags are not always intentionally shared. As I explore below, cross-use of hashtags between different communities on Tumblr can cause confusion for the user.
2. While research on how quilters are using hashtags and Tumblr generally is a new direction for my scholarship, I have been working on quiltmakers communicating through other social media channels since 2008. This study of the use of hashtags by quiltmakers and other crafters is an offshoot of research that I did for my dissertation, *Fiberspace* (2011), in which I looked at the intersection of textiles and technology online, especially how quiltmakers use Facebook. In addition, my essay in *Writing History in a Digital Age* (2013), "Citizen Scholars: Facebook and the Co-Creation of Knowledge," looks at how lay quilt historians use Facebook for scholarly communication.
3. Several fandoms and gamer communities have adopted the practice of calling a guilty pleasure a "quilty pleasure." For example, in the *League of Legends* thread on Mmo-Champion.com, a user writes, "The first time 9gag has given me a chuckle (it's my quilty pleasure)" (http://www.mmo-champion.com/threads/917321-The-League-of-Legends-thread?p=31907507). On Tumblr, the practice seems to be done primarily by fans of Zac Efron. The reason for this is unclear.
4. While the word "sewist" is not yet in the *Oxford English Dictionary*, it is in common use by people who sew. The word has a more contemporary feel than the correct term, "sewer," and cannot be confused with a place where sewage goes.

REFERENCES

Arora-Kayd. (2014, June 3). Attempting to make a quilt! If the blocks come out anywhere close to square I'll count it as a win XD. Retrieved from http://arora-kayd.tumblr.com/post/87735031871

Brackman, B. (1993). *Encyclopedia of pieced quilt patterns*. Paducah, KY: American Quilter's Society.

Dougherty, J., & Nawrotzki, K. (2013). *Writing history in the digital age*. Ann Arbor: University of Michigan Press.

Finken, T. (2014, Spring). 42 Hashtags. *Fons & Porter's Love of quilting*. Retrieved from http://www.fonsandporter.com/articles/42-hashtags-scrap-quilt-project

Heim, J. (1995). *The needlecrafter's computer companion: How to use your computer for sewing, quilting, and other needlecrafts*. San Francisco: No Starch Press.

Heim, J., & Hansen, G. (1998). *The quilter's computer companion*. San Francisco: No Starch Press.

Herman, J. (2013, August 23). Hashtags 101. *Jaybird quilts*. Retrieved from http://www.jaybirdquilts.com/2013/08/hashtags-101.html

Lévy, P. (2001). *Cyberculture*. Minneapolis: University of Minnesota Press.

Libby-on-the-label. (n.d.). Pretty sure I'm going to end up destroying all of the nerve endings in my fingertips soon. Retrieved from http://www.tumblr.com/search/quilt+life

Marrow-j-houndzower. (2013). This is one of the best shows ever and now they are in quilt form for you!!! Retrieved from http://marrow-j-houndzower.tumblr.com/post/61254203796/this-is-one-of-the-best-shows-ever-and-now-they

Miller, S. (2012, October 11). Sewer vs. sewist. *Threads Magazine*. Retrieved from http://www.threadsmagazine.com/item/27517/sewer-vs-sewist

Nadine64. (2014, June 7). Sunbonnet Sue. Retrieved from http://nadine64.tumblr.com/search/quilt

Pixel Quilts. (2014, June 13). Open call for someone who is perhaps overly obsessive with regards to both fandom and quilting because I need advice and someone who won't judge. Retrieved from http://pixelquilts.tumblr.com/post/88659719651/open-call-for-someone-who-is-perhaps-overly

Redfield, R. (1955). *The little community: Viewpoints for the study of a human whole*. Chicago: University of Chicago Press.

Sikarskie, A. G. (2011). *Fiberspace* (Doctoral dissertation). Available from ProQuest Dissertations and Theses database. (UMI No. 3450410)

Sikarskie, A. G. (2013). Citizen scholars. In J. Dougherty & K. Nawrotzki (**Eds.**), *Writing history in the digital age* (pp. 216–221). Ann Arbor: University of Michigan Press.

Tabor, K. (2011, November 29). How-to Twitter for crafters: All about hashtags. *The Nerdy Sewist*. Retrieved from https://nerdysewist.wordpress.com/category/twitter-for-crafters/

Thetwotwoone. (2014a, May 30). What the filth flarn filth am I doing here? Retrieved from http://thetwotwoone.tumblr.com/post/87365233530/what-the-filth-flarn-filth-am-i-doing-here

Thetwotwoone. (2014b, May 30). Where it all began...Retrieved from http://thetwotwoone.tumblr.com/search/where+it+all+began

5sos-stole-my-underwear. (2014). When im old im gonna turn all my band shirts into quilts and use them as bedding for my children. Retrieved from http://5sos-stole-my-underwear.tumblr.com/post/84587165047/this-couldnt-get-any-more-accurate

Jokin' in the First World: Appropriate Incongruity and the #firstworldproblems Controversy

ANDREW PECK

I'm honestly bothered when people text me that don't have an iPhone
#FirstWorldProblems
—StephanieBrunoo, December 5, 2012

A young Haitian boy leans wearily against a tree. Behind him, a hog laps up standing water from a mud puddle at the base of a crumbling stone wall. A rooster struts past discarded refuse and fallen, decaying bricks. The camera pans left and focuses on the boy; his hair is cut short, and he wears a loose-fitting red shirt. Not yet in his teens, he looks small against the trunk of the large palm tree on which he rests his slim frame. Looking at the camera, his eyes begin to wander and travel over the distant ground as he speaks. The voice is his; the words are not. "I hate it," he says, "when I tell them no pickles and they still give me pickles."

This video, only a minute in length, features nearly a dozen Haitians in similar circumstances reading tweets that use the hashtag #firstworldproblems.[1] Released in October 2012 by the organization Water is Life, the video attracted significant attention, accruing over 6 million views on YouTube. And, if the disjunction between the frivolous complaints and the impoverished circumstances remains unclear, the advertisement ends with a tagline that removes any ambiguity—"#FirstWorldProblems **are not problems**" (TheGiftOfWater, 2012, emphasis in original).

Reactions were polarized. To some users, the problem with **#firstworld-problems** was that it enabled insensitivity through a veil of irony—tantamount to telling a racist joke and backpedaling with "I was just kidding" when challenged. Many other users, however, were critical of this interpretation of **#first-worldproblems** and accused the video of missing the point. These users defended **#firstworldproblems** as a self-aware joking performance that acknowledged the complete triviality of these issues and—consequently—implied reflection on one's own privilege.[2] In this view, humor invites a reflection on seriousness and can function as a discursive entry point into more serious discussion.

While both sides agree **#firstworldproblems** is intended to be humorous, they disagree as to the social function of that humor. Using this shared expectation for humor as a baseline, this chapter approaches the **#firstworldproblems** controversy from the standpoint of contemporary humor theory. Understanding how the discursive circulation of these textual fragments creates publicity means observing localized exchanges surrounding this humor and attending to variation and change over time (Warner, 2002, p. 420). Hence, this chapter adds to conversations surrounding publics, digital humor, and social media by drawing on the work of humor theorist Elliott Oring in order to suggest that analyzing how specific genres of humor—such as **#firstworldproblems**—function socially entails an attention to how various appropriate incongruities create, frame, and partake in conversations about their subjects.

APPROPRIATE INCONGRUITY

Humor is not monolithic. It can be broken down into specific types (such as parody, satire), and many of those subtypes can be differentiated even further; for example, satire can be playfully Horatian or contemptuously Juvenalian. In recent years, studies of political satire and parody have been especially common (e.g., Baym, 2005; Boler & Turpin, 2008; Gray, Jones, & Thompson, 2009; Hariman, 2007a, 2007b; Jones, 2009; Painter & Hodges, 2010; Warner, 2007). Taken collectively, these works suggest that satire has a productive social function that can create communities and question hegemony.

These scholars are correct about the positive social potential of satire, but I am hesitant to summarily consider **#firstworldproblems** satirical. Preemptively aligning **#firstworldproblems** with satire binds me to a certain set of assumptions about how these communications function. Instead, I turn to humor theory because humor is the most basic frame that defines these communications.

The incongruity theory of humor is the dominant perspective in contemporary humor scholarship (Kuipers, 2008, p. 361). Incongruity theory suggests all humor plays with how symbols are identified both with and against social and cultural norms, and "If there is a discrepancy, the humorist registers the incongruity between the perceived event and the expected norm to find humor in the relationship" (Lynch, 2002, p. 428).

Oring's (2003) work on appropriate incongruity suggests all humor involves the perception (although not necessarily resolution) of "an appropriate relationship between categories that would ordinarily be regarded as incongruous" (p. 1). The intervention Oring is making in humor theory is the idea that these incongruities must be "appropriate" (after all, not all incongruities are humorous). Appropriateness refers to the ways in which connections between the multiple incongruities at play in any given humorous communication are perceived at the individual level—"the recognition of a connection even if that connection is logically or empirically questionable" (2010, p. 2). Appropriateness links the otherwise disparate parts of the joke (the incongruities).

This notion of appropriateness, as Raskin (2011) has noted in a critique of Oring, can be amorphous and idiosyncratic. But, according to Oring (2011), that is the point. More elaborate analytical frameworks often shift the focus from the observation of how individual humorous communications work to how they fit within the schema of that analytical framework. Hence, the implementation of such frameworks tends to privilege and suggest certain methods of interpretation. As Oring puts it:

> [A]ll humor depends upon the perception of an incongruity that can nevertheless be seen as somehow appropriate. If these terms seem vague, it is because they need to capture what is going on in the joke without any pre-commitment to the categories of a formal theory.... The attempt to impose an abstract template on a corpus of jokes may cloud the process of analysis. Once it is felt that one knows what parts a joke should have, the identification of these parts can become a rote procedure with the jokes being plugged into pre-established categories. The analyst is insulated from the properties of the joke by the categories and terminology. What gets analyzed is no longer the joke but the terminology and categories of the template into which the joke has been cast. (2011, pp. 213, 219)

Oring's appropriate incongruity theory is valuable not only for its definitional aspects but also for the ways in which it encourages direct observation and interpretation in order to understand how jokes function socially and create meaning.

This focus on the interpersonal aspect of humor is key because humor is a quintessentially social phenomenon. As Giselinde Kuipers (2008) adroitly puts it: "[H]umorous utterances are socially and culturally shaped, and often quite particular to a specific time and place. And the topics and themes people joke about are generally central to the social, cultural, and moral order of a society or a social group" (p. 361). This suggests a mutually constitutive relationship between appropriate incongruities and sociocultural contexts. Understanding how relationships between incongruous concepts are rendered appropriate means observing how these jokes are deployed, discussed, debated, and delivered. This means humor is fundamentally discursive.

Observing humor as discursive is necessary because the social function of humor is often spatially and temporally contingent. This contingent nature is reflected in the vast array of different scholarly explanations as to the functions of humor. It has been suggested that humor may function to: maintain social order, express conflict, release social tensions, construct social relations, provide an alternative conception of the world, make sense of everyday life, act as a form of aggression, and provide a form of transgression (Kuipers, 2008). Despite this dizzying list of functions, the most important aspect of these approaches is the underlying suggestion that humor says as much about the teller as it does about the target.

Positioning humor as discursive means analyzing the cultural role of the joke's subject(s) with how specific humorous communications frame and comment on these categories through appropriate incongruities. "These oppositions," Oring (2011) writes, "bear on the potential *meanings* of the jokes to the people who tell and hear them" (p. 219, emphasis in original). Understanding how these humorous communications work together to create meaning as part of a larger discourse means striving for a better understanding of ourselves. Oring (2003) puts it best: "Joke cycles are not really about particular groups that are ostensibly their targets. These groups serve merely as signifiers that hold together a discourse on certain ideas and values that are of current concern" (p. 65).

METHOD

The cases discussed below are derived from qualitative research conducted between October 2012 and December 2013. During this time, I observed hundreds of tweets and dozens of conversations using the **#firstworldproblems** hashtag. Noting its sharp rise in popularity after 2011, I also gathered several pre-2011 examples in order to explore how mainstream adoption of the hashtag on Twitter may have changed the focus of the humor and related conversations over time. The examples presented here are not exhaustive; they are meant to illustrate several observed trends.

JOKIN' IN THE FIRST WORLD

The precise origin of **#firstworldproblems** is nebulous. One of the earliest popular examples emerged from a 2009 discussion thread on the SomethingAwful.com forums titled simply "**#Firstworldproblems**" (Fortune Favors Diebold, 2009). The thread's humor was positioned not as mocking an already existing hashtag, but as a general reaction to annoyingly banal and self-important tweets, as well as the then-popular "FML" ("fuck my life") hashtag and website. The thread-opening post provided other users with an initial joke: "I had to walk 15 minutes to get to the car today. #firstworldproblems." This original joke, as well as those that followed, was rendered appropriate by referencing the social conventions of Twitter while also highlighting the extreme frivolity of the problem by adding the incongruously self-aware hashtag.

Subsequent posts began to stretch the joke's critical bent. Whereas many posters created humor through banality, others began engaging in social critique of both the First World and the notion that problems in the First World are necessarily frivolous. In doing so, these users added another level to the thread's critical play. For example, user "WORST POSTER EVER" wrote, "i got tied to a fence and beaten to death because i was gay #firstworldproblems." The appropriateness of this remark is manifold, referencing social conventions surrounding tweeting, posting in the #firstworldproblems thread, and seeing the First World as civilized and unproblematic. The incongruity in this comment stems from the way this user violates the expectations of these conventions and, in doing so, uses humor for social critique. It not only draws attention to frivolity on Twitter but also critiques the thread for an unproblematic take-up of "First World" as a base term for critical humor.

Several users followed this example and posted similarly critical comments to the thread. User "MRI chalk" posted, "My government is responsible for incomprehensible atrocities of unmatched scope and my society is guilty of pardoning them #firstworldproblems." Another user sarcastically asked if the American Midwest counted as First World. When one user questioned the often disparate hetero/homosexual marriage laws in the First World, another replied that "i'm just saying its probably tough to get riled up about marriage rights when you're starving." User "story" called out the thread for its problematic language, posting that "1st World/3rd World terminology is outdated and obscures the reality of wealth and living standard disparity, hiding the real systematic oppression that millions struggle with #firstworldproblems." Two posts below this, user "Goatstein" wrote, "ugh i still can't believe they canceled Firefly #firstworldproblems."

Despite the seemingly large disparity between these two types of comments—one being aggressively banal and the other using the 140-character format to air major grievances with the Western world—both find common ground in a hyperbolic (and often satirical) critique of Twitter through humor. This humor not only mocked overly self-centered communications by Twitter users (rendering an appropriate baseline) but also characterized these users through implied incongruities between humor and reality. These jokes variously positioned Twitter as a place for the socially unaware or a location for communication preoccupied with the self and individual identity.

The 2009 Something Awful thread was marked by discursive interactions that used humor to critique problematic trends, from terminology to social media to human rights. Although this may not initially seem like much of a discussion, these users engaged with each other in discursive ways—reposting sentiments they agreed with, one-upping jokes made by others, or responding with even more exaggerated forms of humor. This thread was defined by a culture of escalating and responsive humor and critique.

What, then, is the fundamental difference between this comment from the 2009 thread: "I'm still just posting bullshit on the internet to make myself feel better about exploiting the laborers of the third world #firstworldproblems" and this one from Twitter in 2013: "Got a little too ambitious with the burrito I made for lunch, and I had to eat it with a knife and fork. #firstworldproblems" (Harding, 2013)? While the former represents a self-aware type of cultural critique, the second is seemingly a complaint that performs the construction and maintenance of self and group identity. And, problematically, the second type has become increasingly common over the last several years.

On any given day, a search of #firstworldproblems tweets returns results such as:

I love playing golf...but I'm allergic to fresh cut grass.... #FirstWorldProblems (Mike Mc, 2013)

Starbucks ought to have a speed line for those who want coffee instead of milkshakes. #firstworldproblems (Spencer, 2013)

Needing to take my iPhone 5 charger w/me everywhere I go really defeats the purpose of having a "lighter, slimmer" phone. #firstworldproblems (Barakat, 2013)

Whereas the Something Awful thread mocked communications that were self-serving and overly banal, these tweets tend to be more about conveying the personal identity of each user to that user's community of followers. The hashtag seems almost perfunctory—a way of excusing each individual's complaints post hoc. Instead of finding incongruity between what is considered appropriate for Twitter and the greater problems at play in the world, these tweets are more evocative of playing on a perceived incongruity between the First and Third Worlds. In short, the appropriate incongruities at play appear to have changed over time.

But perhaps this is an unfair characterization. After all, most of the 2009 comments on Something Awful are seen as mocking Twitter only in the context of the thread itself. It may be possible that the 140-character limit, as joked about in the Something Awful thread, inhibits complex messages. It bears exploring, then, how these humorous communications play out in digital locations that are more conducive to discussion in order to determine whether the hashtag or Twitter is enabling these messages.

Unfortunately, the large-scale appropriation of **#firstworldproblems** seems to have diminished the humor's critical potential in many locations on the Web. This is true not only on Twitter but also on websites with forums or discussion threads dedicated to **#firstworldproblems**. The tone of current interactions on websites such as Reddit or Something Awful mirrors the problems in the tweeted examples above.

Take, for example, several discussion threads on Reddit's **#firstworldproblems** subreddit. Here users post on topics such as "I was playing on my phone before my food came out, and now that it's here I look like I'm trying to take food pics like a hipster," "I want to check out some cool restaurants in the area but I never get the chance because I get free catered dinner at work," and "I moved, so all of my Chrome autofill information is incorrect."

Unlike the satirical posts from 2009, these uses of **#firstworldproblems** often led to serious discussions on how to tackle the inconsequential issues facing the original poster. The conversations surrounding a thread titled "The Wi-Fi at work is really slow today, but I can't complain to anyone because I'm not supposed to know the password" involved such helpful tips as "Go to the basement of your office building and find the main power cutoff....Sometimes the router just needs to be restarted" (followed by a 20-post discussion on why this is a poor suggestion), "Send an anonymous email to IT" (with a link on how to do so), and "There was a solar flare yesteray, i think it effected internet speeds, mine was slow too yesterday" (literacolax, 2013).

In these discussions, the original poster uses the guise of **#firstworldproblems** to either ask for help for an inconsequential problem or perform an act of self and group identity. Another thread titled "My sandwich costs $5.83, but I only have a $5 in my wallet so I have to use my card" emerged similarly, with the original poster using the thread not only as an excuse to vent his frustration but also as an opportunity to start a conversation on bank card etiquette that supported his decision to pay with a credit card (BeatleFloyd, 2013). As another user commented, empathizing with the topic creator, "My bank manager had a little mini rage at me for using my debit card for small purchases like this. Fuck you guy, I'll spend my money however I want."

Much like the tweets, the humor here is not about a self-aware addressing of one's own frivolity; it is about the performance of identity which, tacitly, also serves to Other the non-Western nations through terminology it takes up unproblematically. The sincerity of these threads and the discussions they create suggest that the incongruity at play is no longer about the triviality of everyday digital communication (after all, this would lead to much different discussions that would deal with and joke about frivolity); it is about performing identity and group membership as a Westerner.

This becomes evident in the "sandwich" thread when one user calls it out as a specifically American problem, saying that in the U.K., the sandwich would cost an even £5. User "NixonsGhost" suggests simply paying with a card all the time, but when told by user "bonestamp" to remember to pay off the card, NixonsGhost replies, "Paying off? We pay with cash from a card—we aren't some third worlders who need to take out a small loan for every purchase." When bonestamp responds about the benefits of paying with a credit card (such as no fees or an extended warranty), a third user, "honeybradger," takes issue, saying "You sound like a fucking idiot for turning down free money. I think you're the third-worlder."

These comments portray a vision of the world in which those not living in the First World are both always impoverished and incapable of making smart decisions with money. These responses are emblematic of a larger trend that disregards the self-aware joking potential of #firstworldproblems and takes the First/ Third World dichotomy seriously. Judging by the incongruity at play here, this is not surprising. In this thread—like many others—the joke is rendered appropriate because of the suggestion that "modern" conveniences (cell phones, casual spending, ready access to food and the Internet) are inherently absent in the Third World. It creates an identity for "us" in a way that flattens non-Western regional differences into overly simplified and disingenuously represented categories.

CONCLUSION

In this chapter, I have suggested the utility of using Elliott Oring's theory of appropriate incongruity to examine digital humor. This theoretical perspective attends to not only the incongruities at play in a joke but also the underlying assumptions linking those two incongruous categories together. Using this model, I have observed how #firstworldproblems has changed over time from a critique of the frivolity of social media to a performance of self and group identity.

I do not wish to suggest that communication lacking cultural critique is always problematic. After all, there is nothing inherently bad about humor for humor's sake or about discourse that performs self-identity or group maintenance. Nor do I wish to criticize #firstworldproblems based solely on its choice of terminology or by claiming that it directly affects non-Westerners in a negative way. Indeed, this humor reveals more about the tellers (and their social media use) than the targets. What is most problematic here is not that this humor reinforces certain cultural categories; it is the way in which #firstworldproblems has come to be used to accomplish these ends.

The reason #firstworldproblems has become so problematic is not because it Others non-Western groups; #firstworldproblems is problematic because, *despite this process of Othering*, it still thrives under the veneer of self-aware, humorous cultural critique. The hashtag has become antithetical to its original deconstructive intent, and, over time, it has become naturalized and lost much of its original capacity for transformation.[3] The humor here has transformed from being critical to reinforcing the very norms it was criticizing. Instead of functioning as a way to call attention to personal privilege or banal egocentric performances, it enables them. It simultaneously reinstantiates a split-world dichotomy and enables egocentrism while also excusing participants of responsibility, blame, or critical awareness under the guise of a joke.

NOTES

1. "First World" was originally meant to divide the world based on economic systems. Capitalist countries, which were predominantly Western, represented the First World; communist countries and Soviet satellites constituted the Second World; and neutral and nonaligned countries formed the Third World. After the end of the Cold War, these definitions began to shift. "Second World" became an outdated term and "First World"—still aligned with the West—began to take on a definitional valance as representing those countries that were at the forefront of social, technological, and developmental progress. In short and with few exceptions, "First World" has come to signify an implicit system of ranking that privileges Western values as being at the forefront of modernity (see also Hall, 1992; Said, 1978; Shah, 2011).

2. Although it is not the focus of this article, there was also significant backlash against what was perceived as the video's exploitation of poor Haitians for Water is Life's PR campaign. The featured Haitians were reportedly "in on the joke" and found the ad campaign quite humorous.

3. Some users have pushed against this process of routinization through the creation of #firstworldproblems variants, such as the inverted Third World Success or the absurd Seventh World Problems. Although not the focus of this chapter, many of these variants have found varying levels of success or popularity in their attempts to reinvigorate the criticality of this humor by stretching its boundaries.

REFERENCES

Barakat, M. [maybarakat] (2013, May 5). Needing to take my iPhone 5 charger w/me everywhere I go really defeats the purpose of having a "lighter,slimmer" phone.#firstworldproblems [Tweet]. Retrieved from https://twitter.com/maybarakat/status/331092733836009473

Baym, G. (2005). The Daily Show: Discursive integration and the reinvention of political journalism. Political Communication, 22, 259–276.

BeatleFloyd. (2013). My sandwich costs $5.83, but I only have a $5 in my wallet so I have to use my card. Retrieved from http://www.reddit.com/r/firstworldproblems/comments/1ed85i/my_sand

Boler, M., & Turpin, S. (2008). The Daily Show and Crossfire: Satire and sincerity as truth to power. In M. Boler & S. Turpin (Eds.), Digital media and democracy: Tactics in hard times (pp. 383–403). Cambridge, MA: MIT Press.

Fortune Favors Diebold. (2009). *#Firstworldproblems*. Retrieved from: http://forums.somethingawful.com/showthread.php?threadid=3136863&

Gray, J., Jones, J. P., & Thompson, E. (2009). The state of satire, the satire of the state. In J. Gray, J. P. Jones, & E. Thompson (Eds.), *Satire TV: Politics and comedy in the post-network era* (pp. 3–36). New York: New York University Press.

Hall, S. (1996). The West and the rest: Discourse and power. In S. Hall, D. Held, D. Hubert, & K. Thompson (Eds.), *Modernity: An introduction to modern societies* (pp. 184–228). London: Blackwell.

Harding, A. [TheAmberHarding]. (2013, May 30). Got a little too ambitious with the burrito I made for lunch, and I had to eat it with a knife and fork. **#firstworldproblems** [Tweet]. Retrieved from https://twitter.com/TheAmberHarding/status/335165842964750336?lang=en

Hariman, R. (2007a). In defense of Jon Stewart. *Critical Studies in Media Communication, 24*(3), 273–277.

Hariman, R. (2007b). Political parody and public culture. *Quarterly Journal of Speech, 94*(3), 247–272.

Jones, J. P. (2009). *Entertaining politics: Satiric television and political engagement.* Lanham, MD: Rowman & Littlefield.

Kuipers, G. (2008). The sociology of humor. In V. Raskin (Ed.), *The primer of humor research* (pp. 361–396). Berlin: Mouton de Gruyter.

literacolax. (2013). The Wi-Fi at work is really slow today, but I can't complain to anyone because I'm not supposed to know the password. Retrieved from http://www.reddit.com/r/firstworldproblems/comments/1eg75z/the_wifi_at_work_is_really_slow_today_but_i_cant/

Lynch, O. (2002). Humorous communication: Finding a place for humor in communication research. *Communication Theory, 12*(4), 423–445.

Mike Mc [MikeMcPatriot]. (2013, May 30). I love playing golf...but I'm allergic to fresh cut grass.... #FirstWorldProblems [Tweet]. Retrieved from https://twitter.com/MikeMcPatriot/status/340255882090782722

Oring, E. (2003). *Engaging humor.* Urbana: University of Illinois Press.

Oring, E. (2010). *Jokes and their relations.* New Brunswick, NJ: Transaction Publishers.

Oring, E. (2011). Parsing the joke: The general theory of verbal humor and appropriate incongruity. *Humor, 24*(2), 203–222.

Painter, C., & Hodges, L. (2010). Mocking the news: How *The Daily Show with Jon Stewart* holds traditional broadcast news accountable. *Journal of Mass Media Ethics, 25*(4), 257–274.

Raskin, V. (2011). On Oring and GTVH. *Humor, 24*(2), 223–231.

Said, E. W. (1979). *Orientalism* (1st Vintage Books ed.). New York: Vintage Books.

Shah, H. (2011). *The production of modernization: Daniel Lerner, mass media, and the passing of traditional society.* Philadelphia, PA: Temple University Press.

Spencer, J. [spencerideas]. (2013, May 16). Starbucks ought to have a speed line for those who want coffee instead of milkshakes. #firstworldproblems [Tweet]. Retrieved from https://twitter.com/spencerideas/status/335037534159642625

TheGiftofWater. (2012). First world problems anthem. Retrieved from http://www.youtube.com/watch?v=fxyhfiCO_XQ&feature=youtu.be

Warner, J. (2007). Political culture jamming: The dissident humor of *The Daily Show with Jon Stewart. Popular Communication, 5*(1), 17–36.

Warner, M. (2002). Publics and counterpublics (abbreviated version). *Quarterly Journal of Speech, 88*(4), 413–425.

CHAPTER FOURTEEN

#RaiderNation: The Digital and Material Identity and Values of a Superdiverse Fan Community

ANTHONY SANTORO

On May 21, 2013, Charles Woodson signed a 1-year contract with the Oakland Raiders, returning to the team that drafted him, after seven seasons with the Green Bay Packers. Woodson had worn jersey number 24 as a Raider for his first eight seasons and wanted to do so again, but Tracy Porter, whom the Raiders had signed earlier that offseason, then had the number. The "locker room" rules that typically govern such disputes among players failed to resolve the argument; the franchise eventually informed Porter that Woodson would wear 24 (Bair, 2013).

This disagreement among teammates was widely discussed throughout the sports media community and among the Raiders' fans, the Raider Nation. In conversations across a variety of digital media platforms linked via the **#RaiderNation** hashtag, tweeters assessed the situation. Some sided with Woodson, appropriating the popular post-9/11 catchphrase "never forget" and linking to photos of a play in the 2011 playoffs when Seattle's Marshawn Lynch stiff-armed Porter to the ground during his game-clinching touchdown run (http://bit.ly/1oMO7JR). Others appraised the flap in dollars and cents: Woodson jerseys would outsell Porter jerseys, they speculated, thus it is better for Woodson to wear 24 (http://bit.ly/1opIJcv). Others, including a journalist covering the Raiders training camp, continually mentioned Woodson by his Twitter handle—@TwentyFourWines—rather than by name, highlighting the fact that Woodson is branded by his number and by his wine business and declaring him at once producer, consumer, and product—multiply so.

This essay analyzes how distributed conversations tagged **#RaiderNation** link, create, and presence diverse, changing, and yet discrete populations. Loosely following Gavin Milner's lead in treating the hashtag as a discursive knot (Milner, 2013), this essay uses the four lines that form the hashtag as four attributes of these communities: superdiversity, concepts of the civil religious, multiple under-standings of "nation," and engagement with regnant neoliberalism. Beginning with "superdiversity" as a theoretical construct and as it is enacted and embodied via these conversations and participants, the essay then turns to the values the community articulates, in terms of the values implied by self-conscious, ascribed superdiversity, concepts of the civil religious, and the "nation" and how they are articulated and embodied. A substantial part of this articulation and embodiment involves engagement with the neoliberal order—in terms of both participation and regulated confrontation—with which the essay concludes.

DATA

This essay analyzes Twitter data surrounding several major calendrical and life-cy-cle events: Super Bowls 46–48, the 2013 NFL draft, the Raiders' 2013 training camp, and the 2011 death of Raiders owner and NFL icon Al Davis. Using Twit-ter's advanced search feature (https://twitter.com/search-advanced) to search for **#RaiderNation** in each of these date ranges, I selected for "All" tweets, including retweets, and manually coded and analyzed them. Because no search and compi-lation methods can guarantee complete capture of all relevant Twitter data (High-field, Harrington, & Bruns, 2013), this study presumes a substantial subset of that data. The set included over 13,000 tweets, most pertaining to the Oakland Raid-ers; others pertained to homonymically related terms, like the Texas Tech Univer-sity Red Raiders. Using "**#RaiderNation**" as the sole connector helped maintain a discrete data set while leaving open the possibility for expanded searches based on frequently arising connectors such as **#RaiderFamily** or **#blackandsilver**.

THE #RAIDERNATION: A SUPERDIVERSE PRESENCE

In an era "pervaded with *discourses about diversity*," a number of theorists have argued for the need to move beyond traditional markers of diversity—race, eth-nicity, national origin, etc.—and to recognize the world's growing superdiver-sity (Vertovec, 2012, p. 287, emphasis in original). More than an expanded set of metrics by which diversity is measured, superdiversity presumes an increas-ingly fluid conception of multilevel identity formation in response to the increas-ingly fluid exchanges, including changes to migration and work/life patterns, that characterize contemporary life in a global world (Arnaut, 2012). One area where we can see superdiversity enacted is within the "fan nation," an exclusivist identification that is also arguably the most expansive and accommodating canopy under which superdiverse groups can come together in mutual recognition and on equal footing (Anderson, 2011). This occurs not only in the physical world, where sport has long provided ways for outsider groups to negotiate their places in society, but also in the digital, as in the **#RaiderNation** conversations, which we can see in the tweets that followed Al Davis's death.

The only person in NFL history "to serve as a personnel assistant, scout, assistant coach, head coach, general manager, commissioner and team owner," Davis died on October 8, 2011 (Pro Football Hall of Fame, 2014). In a piece on ESPN.com titled "Al Davis Created a Diverse Raider Nation," Bill Williamson highlighted Davis's contribution to breaking the NFL's color and gender lines, hiring the league's first Hispanic and African American head coaches and the first female chief executive (Williamson, 2011). Williamson's representative piece is worth investigating in detail. First, he takes for granted the existence of the "Raider Nation," an idea that is so sufficiently accepted that it can be satirized. The title of an article in *The Onion* declared that the "Latest U.N. Report Shows Raider Nation at Bottom of Human Development Index Rankings" (2013), legitimating the fan nation per se and a number of stereotypes about this particular fan nation, as we will see below. Second, Davis's coaching and executive hires and his treatment of his players are depicted as Davis living up to the ideal declared in Williamson's title—diversity. In the wake of Davis's death, tweets incorporating meaning making into ethnic religious practice (http://bit.ly/1zripog), noting the variety of languages in which Raiders games were broadcast (http://bit.ly/1pXvKUf), or celebrating these hires (http://bit.ly/1tKOMwx) reinforce the idea that diversity was a realized value, not just an ideal.

Tweets following Davis's death displayed a degree of diversity surpassing the traditional, "ethno-focal understanding" of the term (Vertovec, 2007, p. 970). There were certainly ethno-focal tweets, such as one wondering "how many cholos got R.I.P. Al Davis tatted on em this wknd? **#RaiderNation** (pour1out)" (http://bit.ly/1mAZUHd). "Pour1out" orients the tweet, the context, and, via the hashtag, the networked community around the traditionally sacred activity of offering a libation in memory of the departed. The rhetorical question reinforces the importance of memorialization while highlighting the availability of the tattoo to mediate grief and memory. "Cholo," meanwhile, is freighted with negative associations, particularly an implication of gang activity and affiliation, and can be used as a racial epithet (Virgil, 1988, 2012). These associations are part of the way the Raider Nation identifies and is identified. Some of this is by design, as when the community plays up the stereotypes associated with it, but this self-conscious adoption of stereotypes draws on pre-existing notions of alterity that derive from racial, ethnic, and class stereotypes that follow from the team's East Bay origins and the dozen years the franchise spent in Southeast Los Angeles (Cube, 2010; Miller & Mayhew, 2005). Tweets declaring that "Raider fans are the type to give you a [c]oma like how Dodger fans did to a Giants fan" (http://bit.ly/W3g1rH) or that they are trailer park convicts (http://bit.ly/1oKlJqa) affirm these stereotypes and give the community in question a public image to contest and against which to define itself.

The Raider Nation's perception of itself as "outsider" helps explain the emphasis on legitimacy running through these tweets. Emphasizing Davis's importance legitimizes the community. It is also a tactic to make that community legible within the context of a society that they feel tries to render it invisible. Beyond Davis's accomplishments in football, for which he was enshrined in the Pro Football Hall of Fame, commenters linked Davis with other cultural icons, calling him "'The Frank Sinatra of Football' because he did it his way" (http://bit.ly/1qqXcHV). Others linked Davis with Apple's Steve Jobs, who had passed away 3 days earlier, wondering "what God is planning [that] he needs Steve Jobs and Al Davis [in] the same week!" (http://bit.ly/1tuKNrR). Another commented on the "weird synergy" that the "Macintosh [was] introduced during @raiders SB XVIII win" (http://bit.ly/1wa5jtW). These tweets are claims to legitimacy, to a realness, that this community often feels deprived of, in part because of the stereotypes associated with it.

Staking these claims to legitimacy is a manifestation of what Nick Couldry calls "'presencing,'…[the] media-enhanced ways in which individuals, groups and institutions put into circulation information about, and representations of, themselves for the wider purpose of *sustaining a public presence*" (Couldry, 2012, p. 50, emphasis in original). Because this public presence is actually "an aggregate of private spheres" (Miller, 2011, p. 175), looking at the various media used in the presencings within the **#RaiderNation** conversations will help us see the various privates that are mediated and amalgamated into the public. The hashtag, which users forced Twitter to adopt (Siles, 2013, p. 2120) and which allows users to link beyond Twitter to other platforms and media, is the simplest user modification. At the other end of the spectrum of media interfaces is punter Chris Kluwe's use of Google Glass during training camp, offering an "insider's look" via media posted to blogs, Twitter, and YouTube (http://bit.ly/1wa5Dc2; http://bit.ly/1jjn4GL). In between lie various podcasting platforms and apps; mixtures of digital and material media content, such as links to eBay auctions of team-branded smartphone covers (http://bit.ly/U6q9hM; http://bit.ly/1qQmLUU); and an online guestbook hosted by *The Modesto Bee*, which offered mourners a place to voice their grief, light a symbolic candle, or link to sites offering support for those coping with loss. These media gave fans the ability to participate in the developing story while also legitimating their grief by mediating it (Sumiala, 2013).

The links between the digital and "offline" worlds are critical to understanding the articulated private spheres that make up the presenced public—in this case, a public presence that is consumerist, has a strong and somewhat ambivalent theistic component, and capably compresses space-times in its presentation of itself. The consumerist element is most visible in the way the tweeters digitally display how they "rep" the Raider Nation, a term most frequently meaning wearing team apparel in various contexts. Apparel is used to declare a "real" gender or ethnic identity (http://bit.ly/1kd6Udc; http://bit.ly/1opONBX; http://bit.ly/U6qsJv), albeit through a highly commercialized, ambiguous slogan like "Real X Wear Black" (http://bit.ly/1jjnRHK). Apparel in the workplace is also used to legitimate the fans' grief by bringing that apparel into that space and time, an effect most pronounced in tweets from clergy declaring that they either do or will rep their gear while performing their duties—"Super bowl jersey day at church" (http://bit.ly/1mnmp6R)—if only because of the juxtaposition of the sporting signifier with the transcendent (http://bit.ly/1mG0qY9; http://bit.ly/1mRl0qp). Photos showing players who were part of Super Bowl–winning teams allowing fans to pose wearing that player's championship rings (http://bit.ly/1q3t5KS), meanwhile, condense discrete space-times into a material object that is "shared," photographed, and shared digitally. When shared online, this photo presences the event and the variable but discrete condensation of space-times it materializes.

When we unpack the mediated presencings linked via the **#RaiderNation** hashtag, we see a wide variety of private spheres distilled into those public presences. They are, first, "me-forming" practices within the typology of the Foucauldian project of the self (Foucault, 1997; Papacharissi, 2012, p. 1989). They also condense discrete space-times and histories into an online presence; the fans posing with a player's Super Bowl ring merge, at a minimum, the histories of football, the franchise, the Super Bowl, and the player into a single photograph. The fan nation's awareness of its superdiversity is similarly broad—a nonexclusive list of identity markers contained within these tweets would include ethnicity, race, gender, occupation, education level, immigration and citizenship status, and religion. Deployed in order to counteract perceptions of alterity, this embrace of superdiversity is at once consumed and embodied and limited by that consumption; a claim to legitimacy on multiple levels; and an articulation of values that the community sees as ideal and as realized, at times incompletely.

VALUES EXEMPLARY AND INTEGRAL

The **#RaiderNation** conversations show a community that understands itself as superdiverse and celebrates that superdiversity across multiple platforms and media formats. We should read assertions that "At this moment in time, across

every faith, race, country & creed **#RaiderNation** is united in joy" (http://bit.
ly/1m08VZM) as the community's social imaginary, the way it imagines its social
existence, its expectations regarding social interaction, and the taken-for-granted
"normative notions and images that underlie these expectations" (Taylor, 2007,
p. 23). Given the fans' investment in asserting the community's superdiversity,
following Taylor, we can see that it is taken for granted within the imagined
community as distinct from the broader society around them, within which it is
perceived as singular to the point of being one of the community's defining characteristics. Without assessing the validity of this perception, we can see how these
constitutive imaginaries come to be taken for granted. As one tweeter commented,
"when i randomly chose a favorite team in the nfl i never knew it would become
so real" (http://bit.ly/1rdNfik). This recalls Gary Laderman's explanation of the
moral communities that fans create around sports, which, he says, can take on religious significance and "create shared experiences and memories as impressive and
meaningful as any other sacred encounters in this life" (Laderman, 2009, p. 62).
This is the case in tweets proclaiming that Al is "looking down on the Raiders"
(http://bit.ly/1nngz0E), or that the fan would "see you up there my dude" (http://
bit.ly/1m9WMlh). It is also true in cases where commenters explicitly recast Davis
and the Raiders into an allegory.

In an essay published on October 10, 2011, Rabbi Joshua Hess pondered "The
Meaning behind Al Davis' Death on Yom Kippur" (Hess, 2011). Because "traditional Judaism doesn't believe in coincidences," Hess wrote, the fact that Davis
died on the holiest day of the Jewish calendar provided the opportunity to reflect
on why he "was returned to his maker on that auspicious day." The answer, Hess
asserts, can be found in one of the Raiders' mottoes: "Just win, baby." We could
read this as "win at any costs," but Hess argues for a deeper meaning to the phrase.
The road to success is broken and uneven, but no matter how often we stumble,
we should "get back up, dust ourselves off, and keep our eyes on the prize." The
holiday, Hess says, has the same meaning: "forget about the mistakes that you've
made, just try again, and this time, 'Just win, baby.'" Hess continues, "On a day
when God grants us atonement for our mistakes and encourages us to make the
necessary changes to succeed in the future, God brought home one of his emissar-
ies who taught this lesson to…millions…around the world."

Hess's essay inverts Laderman's idea, using an imprecise, even banal catchphrase to frame his theological reflections. But the motto's banality is no barrier to
it being used to drive the community's social imaginary; the banality, multivocality,
and ambiguity of such ideas make them particularly well suited to this construc-
tive deployment (Vertovec, 2012). As one mourner put it, "'Just Win, Baby' 'Pride
& Poise' 'Commitment to Excellence' More than slogans—they're a way of life.
#RaiderNation lives on!" (http://bit.ly/1rdNoSP).

These mottoes have been incorporated into the community's identity and imaginary to the point of functioning as what Tariq Ramadan refers to as a "universal," an answer to the response that he sees to our incomplete understanding of the true significance of our limited, relative viewpoints (Ramadan, 2011, pp. 20–25). The Raiders provide the Raider Nation this ambiguous, multivocal universal. As rapper MC Hammer tweeted following Davis's death, "The Raiders became our flag in the ground" (http://bit.ly/1mG4fN7). This conceit—the flag in the ground—links this articulation of the Raiders' importance to the community to the American civil religion, particularly Will Herberg's notion of the American civil religion as the sanctification of the American way of life. Herberg saw this American civil religion operating as the faith traditions did, "with its creed, cult, code, and community" (Herberg, 1974, p. 76). This civil religion, he says, "provides the framework in terms of which the crucial values of American existence are couched," and through it, national values become redemptive and national history, teleological.

Among the civil religious values that Herberg enumerates are egalitarianism, pragmatism, moralism, individualism, and theism. We can see all of these in the conversations and media linked via #RaiderNation. The groundbreaking hires that Williamson mentions tie with the slogan that Hess explicates—"Just Win, Baby." In this reading, Davis is pragmatic, egalitarian, and moralistic: if he saw an athlete, coach, or executive capable of helping his franchise succeed, he would utilize them. This may seem commonsensical, but the history of football (and other sports) demonstrates that it is not so (Eitzen & Sanford, 1975; Riess, 2014), which tinges such pragmatic rejections of the contemporary racial ethos with a retrospective moralism. One tweeter seamed these threads together in a memorial tweet: "RIP Al Davis. Probably sitting up in heaven right now scouting the fastest, blackest angel. #RaiderNation" (http://bit.ly/1ztDEsh). Here we see civil religious values idealized and declared, realized, and combined with the values put forward by superdiversity, bringing two of the four lines of the hashtag into focus as defined by the community it links: #RaiderNation.

Such comments link these values with the community's self-understanding at local, national, and global levels. Petitions "to keep the @Raiders in Oakland" link the community into a particular local identity by asserting the need to reconcile, or to keep reconciled, conceptions of "home" that may not be compatible, that may in fact be in conflict (Kraszewski, 2008). Research into the economic impacts of publicly financed stadiums has consistently shown that the financial benefits to hosting a professional team are far outweighed by the costs (Delaney & Eckstein, 2003; Ingham & McDonald, 2003). That said, there are significant perceived benefits to having professional sports teams, including the accrual of positive status to the locality, the creation of new forms of community, and the adaptable "universal" that can provide a unifying orientation among a superdiverse population with otherwise conflicting interests—economic, political, and otherwise (Lipsitz, 2002).

"Local" is a problematic designation, however, particularly when used in connection with a billion-dollar global entertainment brand. Sports fandom allows "displaced populations to negotiate home and home identities" (Kraszewski, 2008, p. 140), to (re)construct the local in an increasingly commercialized, mobile society where formerly stable social and community structures have given way to fluid, temporary replacements (Castells, 2004). Sporting universals construct and are constructed by the national, defined both in terms of the nation-state and the community that imagines itself as a nation. To the extent that the Raider Nation's values are taken to be American values—as with superdiversity and civil religion—the community discursively defines itself as prototypically American. To the extent that there is a perceived divergence between the successful realizations of these values within the community as opposed to the nation-state, the community sets itself up as an exemplar to be emulated more broadly by and within the nation-state. The idea of the "one nation" (http://bit.ly/1jHEE7Z) represents the universal, the "flag in the ground," and can stand for the Raider Nation or the United States. As with superdiversity and the civil religious, this emphasis on the nation, be it fan nation or nation-state, gives us our third line in the hashtag: **#Raider*Nation***. As these three lines in these conversations demonstrate, these conversations are intricate assertions of identity and values advocacy via a consumer/entertainment product-cum-universal. This universal, in turn, gives us the means by which to see how the imagined community unites digital and material spaces in its use of neoliberal tools and values in its ambiguous contestation of the ends to which neoliberalism leads.

BUYING AND PRESENCING

The universal that works at the local and national levels also works at the global level. This should not be surprising, given the increasing permeability of national borders via globalization, the international growth of American football (Inoue, Seifried, & Matsumoto, 2010; Maguire, 2011), and the ability of sport to articulate identities. When the Raiders drafted Menelik Watson, a player from Manchester, U.K., tweeters noted that "'[O]akland' is now trending in **#Manchester**" and posted photographs of mergers of Raider and U.K. national symbols, such as a Union Jack hat with a Raider shield at its center (http://bit.ly/1wadvKT; http://bit.ly/1nelUaH). Similarly, the Black Hole Chapter Mexico's Facebook page blends Raider iconography with the Mexican flag in its profile picture (http://on.fb.me/W3oo6x), and when Liberia-born linebacker Sio Moore asked the **#RaiderNation** where they were tweeting from, he received hundreds of replies from all around the world (http://bit.ly/1nkNUhy)—fittingly for a fan base that has declared the Raiders the world's team (http://bit.ly/1yg4nXQ). This global awareness also has a missionistic element: as one fan club described it, their mission is "to help human-kind through a better understanding of the Greatness of the Raiders" (WFIB, n.d.), an element that echoes the civil religious components of the **#RaiderNation** conversations.

Given the community's perception of its superdiversity, we would expect to see that the "'superdiverse' subject may be neoliberalism's ideal multifaceted, 360° con-sumer" (Arnaut & Spotti, 2014, p. 7). The Raiders as a universal, as defined within the **#RaiderNation**, "sells us a story about ourselves" (Lofton, 2011, p. 2)—a story that can be individualized via the team's slogans: Commitment to Excellence, Pride and Poise, and Just Win, Baby. This universal has a double valence: it is made by the very thing that it makes; it engenders affect at the same time that it is engendered by that affect. The universal offers access to betterment through being more "real," but that betterment accrues to always already instantiated elements of identity. The idea that "real women" are Raiders fans, for example, both engenders and is engendered by an extant idea of "real women"; the generative effect is cycli-cal rather than uni- or bidirectional. In selling us the story of ourselves and giving us the means to embody the multiple aspects of our identities, the product gives access both to what we are and to what we can be.

This customizability of identity via the right use of products mirrors the ther-apeutic ethos, the idea that individuals can and should seek to actualize and define themselves via a set of adaptable products and practices that help individuals real-ize their fullest selves. These practices and discourses directed at self-fulfillment "place happiness…solely within the agency of an individual and thereby dovetail a neoliberal discourse that naturalizes the idea of individual autonomy and simul-taneously conceals the supra-individual forces of the social and material world" (Rakow, 2013, p. 486). Sport offers the same possibilities for fulfillment and con-cealment by blending the material and the social at the individual level. Here the individual can be a part of a community that is local, national, and global; is fixed around its universal; and signifies values around which individuals can orient their lives. The emphasis on values, celebrations of superdiversity and inclusion, and the engagement with the national civil religious all underscore Laderman's observation that the entertainment product can mediate collective and individual identities that can be as meaningful as any other allegiances (Laderman, 2009).

At the same time, however, as discussed above, major professional sport exacts a financial price from the communities that provide the stadiums. This is a perfect example of the neoliberal ethos, where a declared public good—the stadium—is publicly financed and then privatized. While taxpayers fund the construction and maintenance of the stadiums, team owners are given favorable lease terms designed not to recoup the expenditure but to prevent franchise relocation (deMause &

Cagan, 2008). Major professional sport reveals the mix of the utopian and destructive elements of neoliberalism (Brown, 2006; Harvey, 2007; Newman & Giardina, 2010). The idealized (and, arguably, realized) superdiversity reflected in the fan community in both the digital and "real" worlds does go a long way toward effacing differences among race, ethnicity, class, gender, national origin, occupation, and education level, at least within certain space-times, as the conversations across various platforms linked via **#RaiderNation** show. These liberatory space-times are in some ways the realization of the Bakhtinian carnival, which allows people freely to test the limits they feel imposed on them outside of those spaces and times (Bakhtin, 2009). They do so, however, via the total commodification of those space-times, which offers freedom from racial, ethnic, gender, and class norms only within the space-times circumscribed by the universal, by the consumer product (cf. Crawford, 2004; Santoro, 2015). The same may be true of the **#RaiderNation** conversations, which likewise enjoy this carnivalesque freedom but are not in themselves able to challenge the norms of mediatized communication (Nelson & Hull, 2008).

Newman and Giardina describe NASCAR as the paradigmatically neoliberal sport, with its "revenues coming primarily in the forms of (a) television rights from broadcasters eager to acquire highly-sought-after racing content, (b) corporate sponsorship deals, (c) ticket sales to live events, and (d) merchandising of stock car–related intellectual properties" (Newman & Giardina, 2010, pp. 1517–1518). The same is true of American professional football, and for the same reasons, yet it paradoxically provides at once an embodiment of the hegemonic structures against which subsets of the population contest, the means for contesting those structures, and the space-time approved for such contestations. It is its own vicious circle of problem and solution, all contained within the material and digital worlds linked by the brands that represent both parts of that circle—problem and solution—working for and "against the individuals they always already interpellate" (Newman & Giardina, 2010, p. 1524).

CONCLUSION

This essay has argued that the concerns of scholars of sport and of new media merge at the point of wondering whether their objects are fundamentally democratic or fundamentally hegemonic. The answer is unclear, but the contours of the question are increasingly apparent. Sport is more than a game; it is an institution, and the people who make up that institution at its various levels are capable of filling perceived "values vacuums" (Rowe, 2003). The values with which sport and sport fan cultures fill these perceived vacuums are frequently drawn from the dominant neoliberal ethos, one that privileges the privatization of public goods, diminishes the import of citizenship in favor of individuals understanding themselves as consumers above all else, and blends with the therapeutic ethos to exhort individuals to find the answers to their own problems—that nothing is beyond their reach with the right combination of right practice, right medium, and right product. The same is broadly true of the new media. Questions about the democratic potential of sport and of digital media remain questions about the extent to which individuals empowered to realize themselves—to inform and "me-form" themselves—are free or determined, at least to the extent to which the practices and products available for such self-realization are on offer. The loop that needs to be investigated more fully, then, is the way these discrete publics purpose the tools of the information age and use them to contest the very neoliberal regimes and ethos that give them both the opportunity and the motive to contest the ends to which these regimes lead, contestations that are generally safely contained within the discrete spaces that the regimes have created and marked out for that very purpose.

REFERENCES

Anderson, E. (2011). *The cosmopolitan canopy: Race and civility in everyday life.* New York: W.W. Norton & Company.

Arnaut, K. (2012). Super-diversity: Elements of an emerging perspective. *Diversities 14*(2), 1–16.

Arnaut, K., & Spotti, M. (2014). Superdiversity discourse. *King's College London: Working papers in urban language & literacies*, paper 122.

Bair, S. (2013, July 24). Raiders give Porter's no. 24 to Woodson [Web log post]. *CSN Bay Area*. Retrieved from http://www.csnbayarea.com/blog/scott-bair/raiders-give-porters-no-24-woodson

Bakhtin, M. M. (2009). *Rabelais and his world.* (H. Iswolsky, Trans.). Bloomington: University of Indiana Press.

Brown, W. (2006). American nightmare: Neoliberalism, neoconservatism, and de-democratization. *Political Theory, 34*(6), 690–714.

Castells, M. (2004). *The power of identity* (2nd ed.). Oxford, UK: Blackwell.

Couldry, N. (2012). *Media, society, world: Social theory and digital media practice.* Cambridge, UK: Polity.

Crawford, G. (2004). *Consuming sport: Fans, sport, and culture.* New York: Routledge.

Cube, I. (Director), & Dhar, A. (Producer). (2010). *Straight outta LA* [DVD]. United States: ESPN Films.

Delaney, K., & Eckstein, R. (2003). *Public dollars, private stadiums.* New Brunswick, NJ: Rutgers University Press.

deMause, N., & Cagan, J. (2008). *Field of schemes: How the great stadium swindle turns public money into private profit* (rev. and expanded ed.). Lincoln: University of Nebraska Press.

Eitzen, D. S., & Sanford, D. C. (1975). The segregation of Blacks by playing position in football: Accident or design? *Social Science Quarterly, 55*, 948–959.

Foucault, M. (1997). Self writing. In P. Rabinow (Ed.), *Essential works of Michel Foucault, 1954–1984* (Vol. 1, pp. 207–222). New York: New Press.

Harvey, D. (2007). Neoliberalism as creative destruction. *The ANNALS of the American Academy of Political and Social Science, 610*, 21–44.

Herberg, W. (1974). America's civil religion: What it is and whence it comes. In R. E. Richey & D. G. Jones (Eds.), *American civil religion* (pp. 76–88). New York: Harper & Row.

Hess, J. (2011, October 10). The meaning behind Al Davis' death on Yom Kippur. *Huffington Post.* Retrieved from http://www.huffingtonpost.com/rabbi-joshua-hess/al-davis-death-yom-kippur_b_1002736.html

Highfield, T., Harrington, S., & Bruns, A. (2013). Twitter as a technology for audiencing and fandom: The #Eurovision phenomenon. *Information, Communication & Society, 16*(3), 315–339.

Ingham, A., & McDonald, M. (2003). Sport and community. In R. C. Wilcox, D. L. Andrews, R. Pitter, & R. L. Irwin (Eds.), *Sporting dystopias: The making and meaning of urban sport cultures* (pp. 17–33). Albany: State University of New York Press.

Inoue, Y., Seifried, C., & Matsumoto, T. (2010). American football development in Japan: A study of National Football League strategies. *International Journal of Sport Management, 11*, 248–271.

Kraszewski, J. (2008). Pittsburgh in Fort Worth: Football bars, sports television, sports fandom, and the management of home. *Journal of Sport and Social Issues, 32*(2), 139–157.

Laderman, G. (2009). *Sacred matters: Celebrity worship, sexual ecstasies, the living dead, and other signs of religious life in the United States.* New York: New Press.

Lipsitz, G. (2002). The silence of the Rams: How St. Louis school children subsidize the Super Bowl champs. In J. Bloom & M. N. Willard (Eds.), *Sports matters: Race, recreation and culture* (pp. 225–245). New York: New York University Press.

Lofton, K. (2011). *Oprah: The gospel of an icon.* Berkeley: University of California Press.

Maguire, J. (2011). The consumption of American football in British society: Networks of interdependencies. *Sport in Society, 14*, 950–964.

Miller, D. (2011). *Tales from Facebook.* Cambridge, UK: Polity.

Miller, J., & Mayhew, K. (2005). *Better to reign in Hell: Inside the Raiders fan empire.* New York: New Press.

Milner, R. M. (2013). Pop polyvocality: Internet memes, public participation, and the Occupy Wall Street movement. *International Journal of Communication, 7*, 2357–2390.

Nelson, M. E., & Hull, G. A. (2008). Self-presentation through multimedia: A Bakhtinian perspective on digital storytelling. In K. Lundby (Ed.), *Digital storytelling, mediatized stories: Self-representations in new media* (pp. 123–141). New York: Peter Lang.

Newman, J. I., & Giardina, M. D. (2010). Neoliberalism's last lap? NASCAR nation and the cultural politics of sport. *American Behavioral Scientist, 53*(10), 1511–1529.

The Onion. (2013, August 20). *Latest U.N. report shows Raider Nation at bottom of human development index rankings.* Retrieved from http://www.theonion.com/articles/latest-un-report-shows-raider-nation-at-bottom-of,33682/

Papacharissi, Z. (2012). Without you, I'm nothing: Performances of the self on Twitter. *International Journal of Communication, 6*, 1989–2006.

Pro Football Hall of Fame. (2014). *Al Davis.* Retrieved from http://www.profootballhof.com/hof/member.aspx?player_id=51.

Rakow, K. (2013). Therapeutic culture and religion in America. *Religion Compass, 7*(11), 485–497.

Ramadan, T. (2011). *On super-diversity.* Rotterdam: Witte de With; Berlin: Sternberg Press.

Riess, S. A. (Ed.). (2014). *A companion to American sport history.* Oxford, UK: Wiley Blackwell.

Rowe, D. (2003). *Sport, culture and the media: The unruly trinity* (2nd ed.). Maidenhead, UK: Open University Press.

Santoro, A. (2015). Unsilent partners: Sports stadiums and their appropriation and use of sacred space. In J. Stievermann, P. Goff, & D. Junker (Eds.), *Religion and the marketplace in the United States* (pp. 240–266). New York: Oxford University Press.

Siles, I. (2013). Inventing Twitter: An iterative approach to new media development. *International Journal of Communication, 7,* 2105–2127.

Sumiala, J. (2013). *Media and ritual: Death, community, and everyday life.* New York: Routledge.

Taylor, C. (2007). *Modern social imaginaries.* Durham, NC: Duke University Press.

Vertovec, S. (2007). Introduction: New directions in the anthropology of migration and multiculturalism. *Ethnic and Racial Studies, 30*(6), 961–978.

Vertovec, S. (2012). "Diversity" and the social imaginary. *European Journal of Sociology, 53*(3), 287–312.

Virgil, J. D. (1988). *Barrio gangs: Street life and identity in Southern California.* Berkeley: University of California Press.

Virgil, J. D. (2012). *From Indians to Chicanos: The dynamics of Mexican-American culture* (3rd ed.). Long Grove, IL: Waveland Press.

Williamson, B. (2011). *Al Davis created a diverse Raider Nation.* Retrieved from http://espn.go.com/blog/afcwest/post/_/id/33406/al-davis-created-a-diverse-raider-nation

Worldwide Fans in Black (WFIB). (n.d.). *Membership.* Retrieved from http://fansinblack.com/x/membership.php

Hashtags in Communities, Polities, and Politics

CHAPTER FIFTEEN

Black Twitter: Building Connection through Cultural Conversation

MEREDITH CLARK

There's power in these Black Twitter streets...motivating masses to do anything creates something. There's always results. A lot of times on Twitter there are just a lot of words, but then something gets done. Someone once told me, "your greatest resources in life are people," and that's especially true of Black Twitter, because we don't do a lot in life to motivate each other. People reach people to get things done. If you reach a lot of people, you can get things done via Twitter.

—@PresidentialHB (personal communication, 2014)

Thanks to the curiosity and extended gaze of mainstream mass media producers, Black Twitter has been definitively framed for its ability to "get things done" through online conversation. Black Twitter forms its own hashtag public through an ongoing process of self- and group-identity maintenance, using hashtagged tweets to set boundaries of inclusion and articulate its values. These users, standouts among the 26% of all African Americans who use the Internet, have often been characterized as a digital mob. Using online messaging to draw attention to news of interest to Black communities in the United States, these users participate in what Brock (2013, p. 529) describes as "cultural conversation" — engaging in the banal, chatting about television shows, and notably, lampooning and lambasting offenders. Their communicative acts contribute to an ever-evolving sense of community (McMillan & Chavis, 1986). Via everyday conversation and the use of hashtags as communication performance, these communicators act and react with one another rather than with an imagined audience, as scholars from Anderson (1986) to Marwick & boyd (2010) have previously described.

Community Through Conversation

The use of so-called "Blacktags," culturally resonant language and phrases combined with hashtags, was cited by academics as one of the differences between Black and non-Black Twitter users. Brendan Meeder, a Ph.D. student at Carnegie Mellon who researched the spread of trends via social networking, explained how these hashtags became so popular (Manjoo, 2010). Meeder explained that close offline relationships and a certain density contribute to the trending ability of Blacktags: "If you have 50 of these people talking about [a Blacktag], think about the number of outsiders who follow at least one of those 50—it's pretty high at that point. So you can actually get a pretty big network effect by having high density" (Manjoo, 2010).

The initial wave of scholarly inquiry into African Americans' Twitter use centered on the use of the hashtag and African American Vernacular English (AAVE) to express a degree of commonality (Brock, 2012; Florini, 2013; Cantey & Robinson, Chapter 16, this volume). Historically, this work is linked to Banks's (2005) analysis of how members use Black orality (spoken AAVE) in their written communication on social forums including the website BlackPlanet. Banks (2005) described how BlackPlanet users drew upon AAVE in their online exchanges, encapsulating the qualities of the spoken word in written form. Byrne confirmed these findings: "[T]hey show how participants can use these traditional communication patterns as markers of cultural and racial authenticity" (Byrne, 2008, p. 320). Yet previous studies on so-called Black Twitter lacked two perspectives: (1) media reports of minority use of computer-mediated communication (CMC) technology and (2) explanation of the phenomenon of "Black Twitter" from the perspective of its contributors. Recognizing Baym's (2006) exhortation that critical cyberculture studies require the researcher to interview communicators in order to make claims about online phenomena, this chapter draws on interview data from 36 research participants to describe Black Twitter's three-level structure, its functions, and its observable impact in social media and news media environments.

The lived experiences of Black Twitter participants contribute to the formation of a self-selecting, constantly shifting, web-based community of Black Internet users. Usually pegged as an outgroup, these communicators use culturally based language, phrases, and references to organize and elevate their group status (Brown, 2000; Tajfel, 1972, 1981; Tajfel & Turner, 1986). Their shared ethnic and cultural background serves as the foundation for what Gruzd, Wellman, and Takhteyev (2011) refer to as personal communities. The connections, built around specific topics of interest that participants repeatedly return to within what Jones (1997) called a virtual settlement, create thematic nodes. As these two levels of community connect through the tweeting and retweeting of hashtagged messages that resonate with Black users, the meta-network of Black Twitter as a whole emerges. Where boundaries of class, education, gender, and geography might otherwise stratify Twitter's Black users, the use of culturally resonant hashtags affords them the opportunity to form multilevel networks online, developing a sense of online community (Blanchard, 2007). Black Twitter as a hashtag public is formed through the uniting of individuals who share some of the interest and characteristics reflective of each participant's physical and virtual identities. As cultural artifacts, the hashtags move through Black Twitter's three levels of connection through a six-stage process of self-selection, identification, performance, affirmation, reaffirmation, and vindication. Based on participant interviews, I've selected two hashtag episodes that my consultants used to describe how Black Twitter's structure and function has contributed to its existence as a phenomenon of cultural communication.

These hashtag conversations are iterative examples of how Black Twitter's structure, grouped into personal communities and interconnected thematic nodes, is used to mobilize and create conversation topics that trend via hashtags. #PaulasBestDishes, a hashtag that trended in the wake of the celebrity chef being sued for workplace bigotry, and #SolidarityIsForWhiteWomen, a hashtag created by Black feminist Mikki Kendall, are two watershed examples of Black Twitter's ability to interrupt mainstream media narratives about Black life online.

#PAULASBESTDISHES: DIGITAL EVIDENCE OF COLLECTIVE SOCIAL IDENTITY MAINTENANCE WORK

One of the first trending topics created by Black Twitter that gathered media buzz emerged on June 19, 2013, when the *National Enquirer* reported that television personality Paula Deen was being sued by a former employee over allegations of racism and bigotry in the workplace. As part of the in-group dialogue, the base phrases used in tweets hashtagged with #PaulasBestDishes consisted of references to Southern food culture and Jim Crow that made sense to other Blacks who had familiarity with the vocabularies in both their original and intended meanings. One example:

> "oh my God. RT @KidFury SDFLHDSKJFADHALK RT@Rebel_Salute: You Hear White Folk Talkin You Better Hushpuppies #PaulasBestDishes. (@crissles, June 19, 2013)

@Rebel_Salute's tweet was retweeted by two Black media elites on Twitter: @Kid Fury (who at the time had more than 70,000 followers) and @Crissles (more than 35,000 followers), hosts of a weekly podcast called *The Read*.

By retweeting @Rebel_Salute's satirical take on the situation, the pair's follower counts gave outsized amplification of the message. Their retweets were indicative of two stages of the process: selection, in that they recognize this particular user, and affirmation, in that they retweet her message to share it with their thousands of followers, along with the comments that they add to the original tweet. @KidFury's addition of what appears to be a series of mistyped characters is the digital syntax for being flustered, flabbergasted, or otherwise amused by the message. His retweet was intentional, as were his keyboard strikes. @Crissles's addition to her retweet was feigned shock. Ultimately, these tweets provided a signal boost to members of the duo's personal communities, which included clusters of other digital media "high centers" who also have tens of thousands of followers (Cha, Benevenuto, Haddadi, & Gummadi, 2012; Wu, Hoffman, Mason, & Watts, 2011). The retweets exponentially echoed the message to dozens, if not hundreds, of thematic nodes and personal communities. All at once, it became an artifact of information and humor, as well as a symbol for others to add their own input. The results of tweeting and retweeting such messages were reflected in a matter of minutes:

> "Uncle Tom's Instant Rice, with butter **#PaulasBestDishes**." (@DebGodFollow June 19, 2013)

> "I thought I was done laughing at the **#PaulasBestDishes** until i saw 'Leggo My Negro Waffles.' I DON DIE!" (@Luvvie, June 19, 2013)

Procedures for Promoting Community Connectivity

These follow-up tweets display several of the six stages of the process of "being Black Twitter." First, the individuals who chose to tweet with the hashtag, adding their satirical offering to the mix through original contributions or retweets, are mostly self-selecting users who are concerned with an issue affecting Black communities. The performance of communicative acts consists of creating tweets around the hashtag's theme; using Deen's own brand to publicly mock and shame her documented bigotry; retweeting with the hashtag to share tweets with their own personal communities and thematic nodes, and commenting on tweets they find particularly incisive or funny, as @Crissles did with a simple "Oh my God," in front of retweeted text. The latter two acts, retweeting and commenting on tweets, are also two examples of affirming other users online. By retweeting, a user is effectively sharing the message with what might otherwise be an untapped audience for the tweet's creator. By commenting, particularly favorably, other users affirm that they have received the original communicator's message, and, with the text of their own comments, either accept it or challenge it. As participants tweeted their contributions to the conversation and retweeted messages emblazoned with the tag, the six-stage process of Being Black Twitter helped form a meta-network of communicators linked by the **#PaulasBestDishes** hashtag. Their connections serve several roles in solidifying Black Twitter as a hashtag public—a phenomenon that has the ability to influence the news day's topic of conversation.

Reshaping an In-group Identity

One participant, @RLM_3, made observations of Black Twitter that point to three potential communicative roles for the active meta-network and its participants: the ability to investigate, uncover and inform; the tendency to employ collective action identity-maintenance strategies to promote social change; and the effective strategy of exposing and publicly shaming the hegemonic in-group's competing social construct of dominance.

> Exposing Paula Deen's racism was under Black Twitter, but some of the things said turned people off. Is this going to be investigative journalism, social activism or name-and-shame? The people who don't consider themselves a part of it is because they get a negative picture of it because of mainstream media coverage. (@RLM_3, 2013)

Such action in the digital space has proven difficult for Black Twitter's out-group to process. For outsiders who have been historically absent or outside the real-world centers of Black cultural conversation, being thrust into the dynamic without a buffer of social courtesy can create a sense of unease. The news media, as @RLM_3 noted, brings this online friction to the attention of a wider audience. The news media's ability to characterize individuals and groups creates communication shortcuts and stereotypes that contribute to negative framing of the phenomenon as it unfolds (Hall, 2003). If the episodes are not interpreted in a culturally competent manner, or if they are solely interpreted through the dominant group's cultural worldview, the resulting texts will lack the context necessary to decode the interactions and their significance to the Black experience, both digitally and in the physical world.

When Black Twitter Strikes

It took a few hours for major news networks to pick up the lawsuit story, and by then, Black Twitter had picked up the pace. It wasn't long before the hashtag's play on words began to trend nationally, prompting coverage by mainstream and alternative media outlets: the trend was covered by Fox News (2013); *Eater*, a special-interest publication for food lovers (2013); BuzzFeed (2013); CNN (2013); and more.

The resulting news coverage is an example of how Black Twitter found vindication after its participants employed the hashtag in a defensive strategy against racism. The creation of a message—all of the shaming tweets hashtagged **#PaulasBestDishes**—became significant enough to warrant news coverage. An additional example of Black Twitter reaching the vindication stage was the release of a statement from Food Network, which said it was "monitoring" the situation on the afternoon of June 19, 2013. Two days after the *National Enquirer* story broke and **#PaulasBestDishes** began to trend, Deen issued three public apologies via personal video messages that were posted to her own website.

This six-stage process of out-group elevation seen in the Paula Deen case demonstrates how Black Twitter users employ humor within their hashtagged messages as an identity-maintenance technique. Participants in that hashtag conversation simultaneously held up some clear examples of how Blacks in the U.S. have been marginalized and discriminated against and joked with others who recognized the symbolism. Participants affirmed the messages being sent in the satiric tweets by retweeting and responding to them, reinforcing the original communicator's message and signaling their company in publicly shaming such microaggressions and racial hostility through creative, humorous means. That the conversations were discussed among my consultants, such as @sherial and her mother, who does not use Twitter, and @RLM_3 and his fellow students at a Midwestern university is another part of the process—affirmation that this interaction and conversation is not just relevant in the virtual realm, but also a part of the community's conversations in the physical world as well.

Finally, the participants who contributed to the hashtag had their sentiments vindicated by the mass media, which made it part of a news item, framing it as an issue of concern to several overlapping interests—centrally, Black consumers, but also entertainment executives, major corporations, and endorsement partners. By selecting the hashtag first and having to rely on tweets as the primary source of indirect input from Black Twitter participants who contributed to the trend, the mass media were able to frame the story for wider audiences, making the central issue of humor as a coping mechanism for dealing with bigotry in the workplace a salient part of the narrative surrounding Deen's deposition. Because traditional news gatekeepers could delve into Twitter to listen in on the cultural conversations without having existing relationships with the communicators, digital communication and raced identity became part of the media narrative, while the complex structures of the communicative network were ignored.

Although this chapter does not establish a causal link between the trends **#Pau-lasBestDishes** and **#PaulaDeenTVshows** and Deen losing both her cable-network contract and endorsement deals, the events and participant narratives suggest that the online phenomenon triggered by Black Twitter's public discourse had impact both in the digital and real worlds. Through the creation, sharing, and retweeting of creative hashtags drawing upon culturally resonant themes, this social media contingent created a wave of negative press with multimillion-dollar implications: "Someone mentioned at the @BWBConference that major brands are afraid of **#BlackTwitter**.... **#TheyDontWantToFaceTheWrath**" (@bgg2wl, 2013).

As Deen's empire crumbled, purportedly in part because of the negative atten- tion she'd garnered in the Twittersphere, the mass media's agenda-setting function was primed (McCombs & Shaw, 1972), and future occurrences of Black Twitter's meta-network mobilization would further advance its selection as a mainstream news item, as evidenced by reaction to the hashtag **#SolidarityIsForWhiteWomen**.

#SOLIDARITYISFORWHITEWOMEN: HASHTAG USE AS SYMBOLIC RESISTANCE

The second episode examined in this chapter is linked to the first by the theme of digital, socially networked resistance to racism. **#SolidarityIsForWhiteWomen** operated on the same assumptions that unite so many participants in Black Twit- ter—a shared experience that includes a historical, systematic marginalization. However, participation in this conversation reached far beyond Black communi- ties in the United States, both online and in the physical world (Guardian, 2013). Its success can be anecdotally linked to connections made within thematic nodes between personal communities of feminists and their allies.

#SolidarityIsForWhiteWomen was created as a response to an ongoing online and offline interaction between Hugo Schwyzer, a White professor, author, and self-proclaimed "male feminist," and several feminists of color, including @Blackamazon, who served as one of my consultants.

In August 2013, retweets from Schwyzer's very public, Twitter-centric melt- down[1] began circulating within feminist circles online, where they eventually attracted the attention of mainstream media outlets (*International Business Times*, 2013). As feminists of color retweeted and commented on Schwyzer's Twitter antics, @Karnythia and @Blackamazon conversed about the deafening silence—in cyberspace and the physical world—of White feminists who had ignored or tried to explain away Schwyzer's self-described "awful" behavior when it was directed toward @Blackamazon and other critics. As the tales of his misdeeds unfolded and were later catalogued in blogs and alternative media, @Blackamazon wondered aloud whether and when her White feminist sisters would come to her aid. They're not coming, @Karnythia reminded her, because **#SolidarityIsForWhiteWomen**. @Karnythia then began to tweet with the hashtag, offering up examples of how White feminists ignored, marginalized, and/or vilified women of color, specifically Black women:

> #SolidarityIsForWhiteWomen when you ignore the culpability of White women in lynching, Jim Crow & in modern day racism. (@Karnythia, August 12, 2013)

The hashtag took off as women around the world used it to discuss slights perpetrated by White feminists against feminists of color. The tweets were not limited to mentions by Black women or women of Hispanic, Asian, and Native American descents. The hashtag was also used by White allies and individuals both male and female to discuss the fault lines of race within progressive community spaces, particularly online:

> #SolidarityIsForWhiteWomen means Rihanna has a responsibility but Miley is just experimenting. (@blogdiva, August 12, 2013)

> #SolidarityisforWhitewomen when pink hair, tattoos, and piercings are "quirky" or "alt" on a White woman but "ghetto" on a black one. (@zblay, August 12, 2013)

> #SolidarityIsForWhiteWomen when you think I need to be saved from the men in my community while ignoring fetishization from the men in yours. (@pushinghoops, August 12, 2013)

> #SolidarityisforWhiteWomen paints #Madonna as a multi-talented feminist icon, while @rihanna & @Beyonce are vapid & hypersexualized. (@weian_fu, August 12, 2013)

> #SolidarityIsForWhiteWomen when convos about gender pay gap ignore that White women earn higher wages than black, Latino and Native men. (@RaniaKhalek, August 12, 2013)

As a hashtag, **#SolidarityIsForWhiteWomen** has had some of the strongest offline transference with respect to conversation topics originating with or being linked to Black Twitter. In the weeks and months after the hashtag appeared, it was covered by mainstream media and adopted by existing social communities as shorthand for discussion of the exclusion of women of color within feminist circles. **#SolidarityIsForWhiteWomen** is a summation of Patricia Hill Collins's description of Black feminism: a demonstration of Black women's emerging power as agents of knowledge (Collins, 1990). Participants in the hashtagged conversation offered personal examples of their experiences being marginalized because of their race, ethnicity, gender, abilities, etc. Some examples include:

When White women are seen as being the default and women of colour are the other / exotic / forbidden **#SolidarityisForWhiteWomen**. (@nursetohbad, August 12, 2013)

#SolidarityIsForWhiteWomen when Lena Dunham is called the voice of a generation even though there are no women of color on "Girls." (@blogdiva, August 12, 2013)

#SolidarityIsForWhiteWomen When Black/brown women get abuse online few even care; when White women get it it's a transnational talking point. (@adnaansajid, August 12, 2013)

#solidarityisforWhitewomen who cry when a woman of color directly confronts their White supremacist and imperialist thinking. (@charlenecac, August 12, 2013)

#SolidarityIsForWhiteWomen calls Hillary the first viable women's candidate even though Shirley was the first and only nominee. (@favstar_pop, August 12, 2013)

These tweets, taken from a cross-section of users, both Black and non-Black, incorporate different voices linked by @Karnythia and @Blackamazon's personal communities and thematic nodes, connecting them to a larger meta-network. On the first levels, many (but not all) of the individuals who tweeted with it were Black.

Cultural Significance and Symbolism of the Hashtag

When I launched the hashtag **#SolidarityIsForWhiteWomen**, I thought it would spark discussion between people impacted by the latest bout of problematic behavior from mainstream White feminists. (@Karnythia, 2013)

Purposefully acting as a Black feminist, @Karnythia's online speech was the digital embodiment of "portray[ing] African American women as self-defined, self-reliant individuals confronting race, gender, and class oppression" (Collins, 1990, p. 221). This hashtag is an example of formal deliberation that Schudson (1997) describes as the type of conversation that occurs in the public sphere with the intention of influencing decision and policy making.

In this instance, @Karnythia acknowledged how feminists of color have been marginalized by their White counterparts, and even when the opportunity to stand in solidarity with the otherwise "weaker" members presented itself, White feminists chose to band together, ignoring or discounting experiences which are unlike their own. The hashtag, as @Karnythia described, was created with an agenda of making feminists of color and the communities they represent visible both within feminist circles and in public conversation on the whole. It evoked a venerated form of protest—the threat of bad publicity—all of which drew greater attention and negative press for the silent majority of feminists complicit in the marginalization of feminists of color.

As a digital artifact that represents some of Black Twitter's boundaries, the **#SolidarityIsForWhiteWomen** hashtag became a symbol that was easily shared between online and physical settings without losing meaning. As a cultural artifact, it was initially embedded with the meaning its creator(s) and initial users ascribed to it. As it grew in popularity and began to trend, a sense of collective identity grew among its users.

#SolidarityIsForWhiteWomen became the inspiration for two similar hashtags that would follow in later months: **#BlackPowerIsForBlackMen** and **#NotYourAsianSidekick**, which were designed to highlight oppression by Black men (many of whom took to Twitter to bash Black women around the **#SolidarityIsForWhiteWomen** hashtag, saying they were being divisive), and the fetishization of Asian Americans. The hashtag lived a divergent existence offline as well. In some cases, it was outright co-opted by pre-existing organizations, without proper attribution of @Karnythia or inclusion of her voice in the panels and discussions it was used to unite and draw attention to. In smaller, more grassroots circles, the hashtag was used as a signifier true to its initial creation—that community support for feminist visibility and activism was limited to privileged (read: White, elite) feminists and shut out feminists of color.

The offline events and media coverage of **#SolidarityIsForWhiteWomen** point to an interesting phenomenon: Black Twitter's participants becoming gatekeepers for information, setting an agenda for mainstream media publications and their consumers. The days, weeks, and months after **#SolidarityIsForWhiteWomen** began to trend, not only was it covered by digital and legacy mainstream and niche media outlets but so was each step of its aftermath. Years later, this hashtag is often cited in discussions about hashtag activism. It has become a widely recognized artifact of digital culture and is arguably a main point of reference for the mass media's framing of the online phenomenon known as Black Twitter: as a "mob" of angry individuals using Twitter—its hashtags, mentions, and retweets—to draw attention to a particular cause.

MAINTAINING BLACK TWITTER'S PUBLIC ORDER

Black Twitter's actions, modeled in episodes characterized by satire, petition, and shaming, have demonstrated that the Black digital presence is one that demands recognition by other users and the mainstream news media. Its individual users, personal communities, and thematic nodes contribute to a greater ability for a linked network of Black communicators to plead their own cause via digital media in a shared space with influence that is quantifiable through follower counts, tweets, and retweets.

Paula Deen was, at best, collateral damage in the phenomenon's growth and power. But her case was a testing ground for the success of individual Black users tapping into their communities to gain visibility around an issue. And through **#SolidarityIsForWhiteWomen**'s metamorphosis from its online origin into offline iterations, Black Twitter incidentally modeled how the phenomenon could be seized upon and replicated by other groups.

To frame Black Twitter as a "mob" is to select specific elements of its presence—simply, the sheer number of participants and what they tweet about—and to ignore the factors of community building through communication and collective action identity maintenance. The communicative acts of these interlinked communities have prompted real-world consequences and lead to the social construction of hashtags as artifacts that carry meaning between the virtual and physical worlds. The essence of this phenomenon is not new; Black bloggers in a linked network had smaller success in earlier years (Pole, 2007). Twitter is simply a new medium for connecting Black communities—personal, intimate ones and those linked by common interests—across physical, economic, and social barriers, giving their members greater agency and visibility.

Questions that arise from reflecting upon the online phenomenon known as Black Twitter are of an interdisciplinary nature and can be explored as inquiry into symbolic communication. Additional studies could be situated in the literature on social movements. Further study stemming might build upon Gates's definition of the process of signifying and advance Florini's (2013) assertions about the truncated language, mixed metaphors, purposeful misspellings, and other linguistic devices used in this form of textual online discourse. Finally, as hashtags are relied upon to organize social movements both online and in the physical world, additional studies of Black Twitter will contribute to the refinement of this framework. This framework is introduced with the intention of including the narratives of individuals within the community under study as a collaborative effort to create meaningful and accurate depictions of a social-media public in the literature.

NOTE

1. Schwyzer tweeted "now you get the truth" in ongoing Twitter posts about his life and career on July 29 and 30. http://www.buzzfeed.com/alisonvingiano/why-did-controversial-feminist-hugo-schwyzer-have-a-twitter#.lcn2K1nQV

REFERENCES

Anderson, B. (1986). *Imagined communities: Reflections on the origin and spread of nationalism.* London: Verso.

Banks, A. J. (2005). *Race, rhetoric, and technology: Searching for higher ground.* Mahwah, NJ: Lawrence Erlbaum.

Barrabi, T. (2013). Hugo Schwyzer Twitter Meltdown: "Male Feminist" Professor Rants About Affair With Porn Star Christina Parreira, Admits To "Fraud". *International Business Times.* Retrieved from http://www.ibtimes.com/hugo-schwyzer-twitter-meltdown-male-feminist-professor-rants-about-affair-porn-star-christina

Baym, N. K. (2006). Finding the quality in qualitative research. In D. Silver & A. Massanari (Eds.), *Critical cyberculture studies* (pp. 79–87). New York: New York University Press.

Blanchard, A. (2007). Developing a sense of virtual community measure. *CyberPsychology & Behavior, 10*(6), 827–830.

Brock, A. (2012). From the blackhand side: Black Twitter as cultural conversation. *Journal of Broadcasting & Electronic Media, 56*(4), 529–549.

Brown, R. (2000). Social identity theory: Past achievements, current problems, and future challenges. *European Journal of Social Psychology, 30*, 745– 778.

Byrne, D. N. (2008). Public discourse, community concerns and civic engagement: Exploring Black social networking traditions on BlackPlanet.com. *Journal of Computer-Mediated Communication, 13*, 319–340.

Cha, M., Benevenuto, F., Haddadi, H., & Gummadi, K. (2012). The world of connections and information flow in Twitter. *IEEE Transaction on Systems, Man, and Cybernetics—Part A: Systems and Humans, 42*(4), 991–998.

Collins, P. H. (1990). *Black feminist thought: Knowledge, consciousness and the politics of empowerment.* New York: Routledge.

Florini, S. (2013, March 7). Tweets, tweeps and signifyin': Communication and cultural performance on "Black Twitter." *Television & New Media.* Retrieved from http://tvn.sagepub.com/content/early/2013/03/07/1527476413480247.abstract

Gruzd, A., Wellman, B., & Takhteyev, Y. (2011). Imagining Twitter as an imagined community. *American Behavioral Scientist, 55*(10), 1294–1318.

Hall, S. (1997). *Representation: Cultural representations and signifying practices.* Thousand Oaks, CA: Sage.

Hall, S. (2003). The whites of their eyes: Racist ideologies and the media. In G. Dines & J. M. Humez (Eds.), *Gender, race, and class in media: A critical reader* (pp. 81–86). Thousand Oaks, CA: Sage.

Jones, Q. (1997). Virtual communities, virtual settlements, and cyber-archeology: A theoretical outline. *Journal for Computer Mediated Communications, 3*(3). Retrieved from http://onlinelibrary.wiley.com/doi/10.1111/j.1083-6101.1997.tb00075.x/abstract

Kendall, M. (2013, August 14). #SolidarityIsForWhiteWomen: Women of color's issue with digital feminism. *The Guardian.* Retrieved from http://www.theguardian.com/commentisfree/2013/aug/14/solidarityisforwhitewomen-hashtag-feminism

Manjoo, F. (2010, August 10). How Black people use Twitter. *Slate.* Retrieved from http://www.slate.com/articles/technology/technology/2010/08/how_black_people_use_twitter.html

Marwick, A., & boyd, d. (2010). I tweet honestly, I tweet passionately: Twitter users, context collapse, and the imagined audience. *New Media & Society, 13*(1), 114–133.

McCombs, M., & Shaw, D. (1972). The agenda-setting function of mass media. *Public Opinion Quarterly, 36*(2). 176–187.

McMillan, D. W., & Chavis, D. M. (1986). Sense of community: Definition and theory. *Journal of Community Psychology, 24*, 315–326.

Pole, A. (2007). Black bloggers and the blogosphere. *The International Journal of Technology, Knowledge, and Society, 2*(6), 9–16.

Schudson, M. (1997). Why conversation is not the soul of democracy. *Critical Studies in Mass Communication, 14*, 297–309.

Tajfel, H. (1972). *Social identity and intergroup behaviour.* Thousand Oaks, CA: Sage.

Tajfel, H. (1981). *Human groups and social categories.* New York: Cambridge University Press.

Tajfel, H., & Turner, J. C. (1986). The social identity theory of intergroup behaviour. In S. Worchel & W. G. Austin (Eds.), *Psychology of intergroup relations* (pp. 7–24). Chicago: Nelson-Hall.

Wu, S., Hoffman, J., Mason, W., & Watts, D. (2011). *Who says what to whom on Twitter.* World Wide Web Committee, 32.

CHAPTER SIXTEEN

#BlackTwitter: Making Waves as a Social Media Subculture

NIA I. CANTEY AND CARA ROBINSON

INTRODUCTION

Methods of negotiating one's identity and challenging spaces on Twitter vary to include challenging religious beliefs, sexuality, politics, and race-related issues. In response to negotiating and asserting one's identity within a space, the hashtag **#BlackTwitter** emerged to spark a keen distinction of identity and values as they relate to race and presence in and on social media. Black Twitter is a medium of exchange for addressing current events, personal interests, social issues, criticisms, and so forth. Consider how, 20 years after O.J. Simpson's Bronco chase, it can now be revisited via Black Twitter discussions in which individuals recount where they were during this time and what feelings emerged while watching the now infamous chase. Examining the presence of Black Twitter through this discourse and notable others (e.g., Trayvon Martin, Juror B37, Sheryl Underwood, and Paula Deen) illuminates the value and meaning of community, a sense of belonging, and, in essence, a space that affirms one's identity, individually and collectively (Cantey, 2010).

In this paper, Black Twitter is defined and discussed as a political, metaphorical, real, or imagined community (Anderson, 2006; Keith & Pile, 1993) that reflects a microcosm of experiences aligned with many interpretations of (as well as discourses of) Black people within the larger context of society. We will argue that the use of Black Twitter represents a subaltern using her voice and claiming space for herself and for her community in a public social space (Cantey, 2010) and that this voice is necessary in advancing African American sociopolitical and socioeconomic issues.

BLACK TWITTER DEFINED

As defined by Hilton (2013, para. 5), "Black Twitter is, loosely speaking, a group of thousands of Black Twitterers (though, to be accurate, not everyone within Black Twitter is Black, and not every Black person on Twitter is in Black Twitter) who a) are interested in issues of race in the news and pop culture and b) tweet a lot." According to the Pew Research Center, 26% of Black Internet users use Twitter (as compared to 14% of Whites and 19% of Hispanics) (Delo, 2013). Moreover, in 2010, Edison Research Group found that 25% of all Twitter users were Black (Bosker, 2011). Twitter's structure and communication channel has clearly become an important part of Black life on the Internet. This is particularly true for Black teens and young adults.

For Black teens and young adults, Black Twitter provides a space akin to an alternative public sphere (Fraser, 1990). As Norris and Odugbemi (2010, p. 10) note, "An effective public sphere depends on opportunities for participation and interaction within civil society." As a result of marginalization, racial, ethnic, religious, geographic, and income groups have created what Fraser (1992) refers to as competing, or counter-, publics. A large, diverse society such as the United States contains many groups needing to compete and contest for power in a negotiated space. The interests served and reflected by competing counterpublics assist members of targeted groups to engage in deliberation and discussion. Access to a space, or sphere, for the creation of identity is crucial. Modern social media, through the diversification of sources, have created an alternative public sphere.

The composition of this alternative public sphere is up for debate. If you are Black and you use Twitter, are you part of Black Twitter? Not according to most definitions. Black Twitter is a specific subset of users who create points of dialogue (i.e., hashtags). Within that user subset, there is a core set of users who serve as the primary "leaders" and content creators. These users, who have thousands of followers, help facilitate points of discussion and debate. This process and structure elicits two points of criticism. First, Black Twitter (and Twitter in general) tends to be dominated by young people; the voices of other Black individuals are lost. This is exacerbated by the fact that there is a core group of individuals driving the discussion. The power of this core group leads to the second critique: Black Twitter, like other online social platforms, is stratified and, despite being accessible to a large number of users, is far from egalitarian.

Discussions, articles, and debates on Black Twitter are common among Black bloggers, journalists, cultural leaders, and other individuals interested in online communication. One common critique of the representations of Black Twitter surrounds its role as representative of the Black community. Black Twitter, like the larger Black community, is diverse. As noted previously, what is often defined as Black Twitter is not inclusive of a large number of Black users of Twitter (Opam, 2013). This debate, among others, makes it necessary to outline the key characteristics of Black Twitter.

Characteristic #1: Black Twitter is a source for online activism and mobilization of the Black community.

From large, national news stories such as the Trayvon Martin case (of which, more below) to strong cultural issues specific to the Black community (e.g., the dialogue surrounding natural hair and hair politics), Black Twitter has demonstrated its capacity to organize a vast number of individuals on matters of importance to its core users. As the role of Twitter has expanded over the past 5 years, it has become a key part of the daily news cycle as journalists, politicians, and other leaders utilize its functions as a communication tool to disseminate and gather information. Black Twitter has harnessed that growing power to shed light on issues that otherwise might have gone unnoticed in the larger society. One specific example is the case of the kidnapping of the Nigerian boarding school students by the terrorist group Boko Haram.

On April 15, 2014, in northern Nigeria, 326 girls (aged 15–18) were kidnapped from their boarding school. In response, Twitter mobilized in an effort to bring awareness and demand action for these kidnapped teenagers (Kristof, 2014). Thus, the hashtag **#BringBackOurGirls** was created. The hashtag quickly spread throughout Twitter, and celebrities from Sean "Puffy" Combs to First Lady Michelle Obama appeared in Twitter pictures displaying the hashtag. The quick mobilization helped garner additional media exposure for the story despite its non-Western location. This campaign provides an example of the power with which Black Twitter can mobilize using hashtags to shed light on issues affecting Black communities and other race-related issues.[1]

Characteristic #2: Black Twitter is a platform for dialogue that discusses issues of particular importance in the Black community.

It is important to distinguish the activism role from the dialogue role. A lot of what takes place on Black Twitter is conversations amongst users. "Black Twitter holds court on pretty much everything from President Barack Obama to the latest TV reality show antics" (Holland, 2014, para. 3). While these conversations have no direct mobilizing role, they often do offer a platform where issues of concern to members of the Black community are discussed and awareness of these concerns is noted in a shared public space. According to Meredith Clark, "We wish to plead our own cause. Too long have others spoken for us" (cited in Holland, 2014, para. 7; see also Clark, Chapter 15, this volume). According to scholars, these conversations also reflect long-held traditions of storytelling in the African American culture (Manjoo, 2010).

Recent hashtag conversations on Black Twitter range from **#dangerousblackkids,** in response to Michael Dunn's conviction for attempted murder in the shooting death of Black teenager Jordan Davis, to **#paulasbestdishes,** in response to celebrity cook Paula Deen's admission to using racial slurs in the past (Holland, 2014; Clark, Chapter 15, this volume). Such conversations have traditionally taken place in the barbershop, beauty salon, church, or even in the high school lunchroom but now are shared online, too. On Twitter these conversations help reinforce the use of online space as an alternative sphere for the sharing and creation of identities.

Characteristic #3: Black Twitter is not a monolith. There is a diversity of voices, which are reflective of the diverse voices in the larger Black community.

Unfortunately, Black Twitter is often characterized as speaking with a singular voice: the voice of Black Americans or of the Black community. Yet Black Twitter is polyvocal. Polyvocal communication is "generated by multiple voices and experiences, with sometimes conflicting interpretations of reality" (Wackwitz & Rakow, 2004, p. 6). Black Twitter is a medium to speak to and from the margins, and the experiences are not always common experiences (2004). In reality, multiple voices are reflected through the lenses of generational differences, various levels of education, and myriad lived experiences which, together, potentially account for conflicting interpretations of Black experience. For instance, opinions on natural hair or hair politics, LeBron James vs. Michael Jordan, and parental discipline methods are just a few examples of these multiple voices and interpretations. Black Twitter is also not just about "Black issues." Black Twitter discusses and debates issues as far apart as the television show *Scandal* and the Boston bombings (or bombers). At the same time, however, it does provide a forum for "Black issues" to avoid balkanization and gain mainstream coverage.

Characteristic #4: Despite the viewpoint diversity on Black Twitter, there is a clear generational divide.

Social media platforms tend to skew toward the young, and Black Twitter is no different. The generational divide on Black Twitter sheds light on two important issues. First, online movements tend to leave out those who either do not participate and/or do not participate in the right ways. Those who are unable to understand the lingo of the Internet or, in some cases, those who make a lot of grammatical errors in their writing find it more difficult to engage. Second, in the case of Black Twitter, those who are "older" who do participate tend to have higher incomes and/or are in positions of power: members of the media, celebrities, academics, and athletes. This brings up concerns regarding the class divide in online communities. While neither of these realities is unique to Black Twitter, it is important to recognize the composition of Black Twitter and the prism through which it speaks.

Characteristic #5: Anonymity is an important component of Black Twitter.

Almost all online communities have at least the option of anonymity. Black Twitter is no different. This anonymity, however, does create a different reality with respect to the discussion of issues specific to the Black community. The discussion of racial issues within society and of general concerns affecting the Black community is often fraught with anger and stereotyping due to the long history of negative representations of Black persons within the larger culture. Black Twitter offers a means through which Black individuals harness power to generate discussion of their concerns within the public sphere. It offers a collective and a place for open, honest discussion rather than a forum where "not stepping on toes" or "keeping one's place" is expected (if not demanded) and which also supports the notion of having anonymity in this shared space as users feel free to share opinions outside the mainstream (opinions they may not feel free to share in other public spaces – online and in person).

Characteristic #6: Black Twitter effectively uses protesting as a means by which to call out behavior that fails to advance equality.

The greatest tool Black Twitter has in its arsenal is the power to protest or make calls to action. There's a rapid response when public figures say or do things that run counter to the ideas of justice and equality, and Black Twitter has pretty much perfected that response, making pariahs out of certain figures and demanding action (Moodie-Mills, 2014).

In our digital culture, individuals are increasingly bound to their words and actions. The online distribution of video or audio (see Donald Sterling—former owner of the Los Angeles Clippers NBA franchise who was recorded, and later distributed, by his ex-mistress stating inflammatory statements about Black fans and players) and the screen-grabbing of tweets are common occurrences. Once something enters the online universe, it is almost impossible for it to disappear completely. As a result, individuals can more easily be held accountable for their words and actions, and Black Twitter has harnessed the power of protest to force responsibility. For instance, Black Twitter helped to fuel objections over Reebok's relationship with rapper Rick Ross, whose lyrics were being criticized as pro-rape. Ross had appeared in an ad for the Reebok Classic sneakers (Associated Press, 2013). Although Rick Ross tweeted an apology—which also created more online discussion—then issued a formal statement, again apologizing, Reebok canceled the deal once protesters showed up outside a Reebok store in New York (2014). In this instance, issues of sexism, misogyny, and the role of hip-hop fluttered Twitter.

Black Twitter creates a space in which issues, words, and actions that promote prejudice, racism, and/or stereotyping of the Black community can be addressed with and to others. Whereas in most public forums it is commonplace for racism and other forms of discrimination to be denied or disputed, the online world makes it clear that racism and prejudice and their associated stereotypes are alive and well in a perceived postracial society.

BLACK TWITTER PRESENCE AND VICTORIES

As the presence of Black Twitter has garnered significant attention over the last 3 years for tackling sociopolitical and socioeconomic issues which affect Black communities, it has made some small victories in activism. For instance, Black Twitter is recognized for increasing the awareness of Trayvon Martin's murder (Clifton, 2014), especially in light of local and national news media absence. During this time, Black Twitter also mobilized youth and other supporters to wear hoodies around their local communities and neighborhoods to show their solidarity and to highlight stereotypical racist and classist perceptions associated with Black youth. Although George Zimmerman was acquitted of all charges, the role of Black Twitter is noteworthy from the announcement of Trayvon's death, during George Zimmerman's trial, and especially after the trial.

Juror B37 was a juror in the George Zimmerman trial. Immediately following the verdict in the case, Juror B37 gave interviews to several media outlets. During those interviews, the juror made inflammatory comments, stating that Trayvon's murder was an "unfortunate incident," that she felt "bad" for George Zimmerman, and that there were riots following the trial (there were not). The juror's comments were viewed as particularly harmful due to the racial stereotyping (e.g., the assumption of rioting) given the context of the case. Following the interviews, it was discovered that Juror B37 had been offered a book deal, and Black Twitter took to Twitter to prevent it. This activism started with one user who targeted the literary agent, and the onslaught of tweets began (NewsOne, 2014). Specifically,

> Genie Lauren (@moreandagain) tracked down contact information for literary agent Sharlene Martin, who was going to represent the juror. Lauren posted Martin's Twitter handle, e-mail address and other information and implored her not to let the juror profit from Trayvon Martin's slaying. She urged her followers to do the same. Scores of people tweeted the agent and 1,343 people signed a Change.org petition in protest. Within hours, the agent and the juror released a statement saying the book deal was off. (McDonald, 2014)

The mobilization was successful, and Juror B37's book deal was rescinded.

Along those same lines, Sheryl Underwood, a Black female comedian and co-host on *The Talk*, was challenged by Black Twitter based on comments she made about Black hair: "'OK, I'm sorry, but why would you save afro hair?' Underwood asked. 'You can't weave afro hair. You never see us at the hair place going look, here, what I need here is, I need those curly, nappy beads. That just seems nasty'" (Sieczkowski, 2013, p. 1). Sieczkowski (2013) further reports, "When Sara Gilbert admitted to keeping some of her own son's hair, Underwood interjected, saying, 'Which is probably some beautiful, long, silky stuff'" (p. 1). Needless to say, Black Twitter erupted, weighing in on Underwood's comments as insensitive, along with questioning her Blackness and describing her comments as a form of self-loathing. Underwood later apologized and noted that her responses were off-mark and that the topic of Black hair is a sensitive one, but also that her Blackness should not be brought into question.

Although the public discussions of Trayvon Martin and George Zimmerman, Juror B37, and Sheryl Underwood are different on the surface (on the topics of wrongful homicide, self-promotion, and hair politics, respectively), all are indicative of a larger struggle for Black identity, voice, and presence in America. Claiming Black identities and voices in America is a process that is often undermined and fraught with racism. Black Twitter represents a space in which one's voice and one's interpretation of identity can be heard—and, to some degree, also seen through their mobilization and activism in the 21st century. Moreover, watching these events unfold in real time reaffirms the role of Black Twitter as a forum for advocacy, activism, and mobilization. Contrary to this notion, detractors (e.g., Murphy, 2013; Rhimes, 2014) argue that Twitter and Black Twitter are far from being media for activism.

For instance, high-profile television show-runner Shonda Rhimes noted in her commencement address at Dartmouth that "a hashtag is not a movement. A hashtag does not make you Dr. King. A hashtag doesn't change anything" (Clifton, 2014). These comments were met with mixed responses from Twitter users. While some agreed with Rhimes, many challenged her position, saying: "To say hashtags aren't activism is the same as saying letter-writing campaigns and phone-a-thons aren't. And you'd be wrong" (@feministajones, 2014); and "Sometimes I wonder what the anti-hashtag crowd would have thought about the folks who made the signs in the Civil Rights Movement" (@fivefifths, 2014). Other critics of Rhimes's statement noted that her own television show *Scandal* is ranked in the top five most viewed shows because of Black Twitter.

Although Rhimes's response to her critics fostered more controversial discussions on the value of Twitter among Twitter users, this discourse also illuminates Black Twitter as a community and a communal space.

BLACK TWITTER AS A COMMUNITY

Black Twitter as a community is a space in which communication of equality is political, explanatory, polyvocal, and transformative (Wackwitz & Rakow, 2004). Consider the above discussions as discussions that are personal and often unheard—and, thus, also political.

> Women's voices are silenced in so many ways that the simple act of speaking may itself become a political act…It is a political act to create and open a feminist floor for debate to explore elements of difference, voice and representation. (Wackwitz & Rakow, 2004, p. 6)

This space is reflective of many silenced voices having an opportunity both to be heard and to explore elements of difference, voice, and representation. Black Twitter is a community which allows an oppressed or marginalized person or group to be heard. In this case, it is a space for the subaltern to speak and be heard even when traditional public spaces are institutional spaces of power privileging the oppressor over the oppressed (Spivak, 1995). In addition, Black Twitter is an explanatory space in which users are able to explain their lived experiences in an effort to provide understanding and clarity with respect to these differences. The goal of informing and educating others about one's struggles and lived experiences facilitates normalizing identities that are publicized as abnormal and different. For instance, the discussion on **#dangerousblackkids** was an exchange highlighting the misrepresentation and interpretation of Black youth in America and how this misrepresentation is used to justify wrongful deaths and killings.

Additionally, Black Twitter is a transformative space. According to Wackwitz and Rackow (2004), to be transformative is to contribute "to intellectual and spiritual growth by providing different perspectives through which to conceptualize experiences and the structures of society" (p. 6). In the spirit of being transformative, Black Twitter allows others outside of their culture to see and potentially understand their experiences. Through this public space, Black Twitter is teaching and educating others on issues significant to the Black community. While Black Twitter is not the only mechanism for advocacy and mobilization, it is a mechanism to promote advocacy, to mobilize, to protest, and to engage in open dialogue in the 21st century.

NOTE

1. On other race-activist hashtags, see Rambukkana, Chapter 2, this volume.

REFERENCES

Ahmed, S. (2006). *Queer phenomenology: Orientations, objects, others.* Durham, NC: Duke University Press.

Anderson, B. (2006). *Imagined communities* (rev. ed.). New York: Verso.

Associated Press. (2013, April 2). Rick Ross apologies for date rape lyrics after Reebok ends rapper's contract. *Daily News.* Retrieved from http://www.nydailynews.com/entertainment/music-arts/rick-ross-apologzies-date-rape-message-u-e-o-n-o-article-1.1315198

Bosker, B. (2011, August 3). Why are African Americans more likely to join Twitter? *Huffington Post.* Retrieved from http://www.huffingtonpost.com/2011/08/02/african-americans-twitter-use_n_916411.html

Cantey, N. I. (2010). *Negotiating space: African descent woman managing multiple marginalized identities* (Doctoral dissertation). Retrieved from ProQuest. (3480335)

Cantey, N. I. (2014). Learning to swim with the barracudas: Negotiating differences in the workplace. In D. Davis-Maye, A. Dale Yarber, & T. E. Perry (Eds.), *What the village gave me: Conceptualizations of womanhood* (pp. 3–18). London: British Library.

Clifton, D. (2014, June 13). What Shonda Rhimes gets wrong about hashtag activism. *Identities. Mic.* Retrieved from http://mic.com/articles/90969/what-shonda-rhimes-getswrong-about-hashtag-activism

Delo, C. (2013, February 14). Pew study: Blacks over-index on Twitter; Whites on Pinterest. *Advertising Age.* Retrieved from http://adage.com/article/digital/pew-study-blacks-index-twitter-whites-pinterest/239810/

Fraser, N. (1990). Rethinking the public sphere: A contribution to critique of actually existing democracy. *Social Text, 25/26,* 56–80.

Fraser, N. (1992). Rethinking the public sphere: A contribution to critique of actually existing democracy. In C. Calhoun (Ed.), *Habermas and the public sphere* (pp. 109–142). Cambridge, MA: MIT Press.

Hilton, S. (2013, July 16). The secret power of Black Twitter. *Buzzfeed.* Retrieved from http://www.buzzfeed.com/shani/the-secret-power-of-black-twitter

Holland, J. J. (2014, March 10). Black Twitter flexing muscles on and offline. *Yahoo News.* http://news.yahoo.com/black-twitter-flexing-muscles-offline-202842249--politics.html

Kalua, F. (2009). Homi Bhabha's third space and African identity. *Journal of African Cultural Studies, 21*(1), 23–32.

Kaya, I. (2005). Identity and space: The case of Turkish Americans. *Geographical Review, 95*(3), 425–440.

Keith, M., & Pile, S. (Eds.). (1993). *Place and the politics of identity.* New York: Routledge.

Kristof, N. (2014, May 3). "Bring back our girls." *New York Times.* Retrieved from http://www.nytimes.com/2014/05/04/opinion/sunday/kristof-bring-back-our-girls.html

Manjoo, F. (2010, August 10). How Black people use Twitter. *Slate.* Retrieved from http://www.slate.com/articles/technology/technology/2010/08/how_black_people_use_twitter.html

McDonald, S. N. (2014, January 20). Black Twitter: A virtual community ready to hashtag out a response to cultural issues. *The Washington Post.* Retrieved from http://www.washingtonpost.com/lifestyle/style/black-twitter-a-virtual-community-ready-to-hashtag-out-a-response-to-cultural-issues/2014/01/20/41ddacf6-7ec5-11e3-9556-4a4bf7bcbd84_story.html

Moodie-Mills, D. (2014, March 14). Will Black Twitter help swing the 2016 presidential election? *The Grio.* Retrieved from http://thegrio.com/2014/03/14/will-black-twitter-help-swing-the-2016-presidential-election/

Murphy, M. (2013, December 18). The trouble with Twitter Feminism. *Feminist Current.* Retrieved from http://feministcurrent.com/8403/the-trouble-with-twitter-feminism/

NewsOne. (2014, February 24). Black Twitter uses social media to power 21st century civil rights movement. *MAJIC 107.5.* Retrieved from http://majicatl.com/2052228/black-twitter-uses-social-media-to-power-21st-century-civil-rights-movement/

Norris, P., & Odugbemi, S. (2010). Evaluating media performance. In P. Norris (Ed.), *Public sentinel: News media and governance reform* (pp. 3–29). Washington, DC: The World Bank.

Opam, K. (2013, September 3). Black Twitter's not just a group—it's a movement. *Salon.* Retrieved from http://www.salon.com/2013/09/03/black_twitters_not_just_a_group_its_a_movement/

Rhimes, S. (2014, June 9). Commencement Address: Dartmouth College. Retrieved from https://www.youtube.com/watch?v=EuHQ6TH60_I

Sieczkowski, C. (2013, September 5). Sheryl Underwood apologizes for comments about "nappy," "nasty" natural black hair. *The Huffington Post.* Retrieved from http://www.huffingtonpost.com/2013/09/05/sheryl-underwood-apologizes_n_3873970.html

Spivak, G. (1995). Can the subaltern speak? In B. Ashcroft, G. Griffiths, & H. Tiffin (Eds.), *The Post–Colonial Studies Reader* (pp. 24–28). London: Routledge.

Tuan, Y. (1977). *Space and place: The perspective of experience.* Minneapolis: University of Minnesota Press.

Wackwitz, A. L., & Rakow, L. F. (Eds.). (2004). *Feminist communication theory.* Thousand Oaks, CA: Sage Publications.

CHAPTER SEVENTEEN

The 1x1 Common: The Role of Instagram's Hashtag in the Development and Maintenance of Feminist Exchange

MAGDALENA OLSZANOWSKI

It's little blurbs we are sharing and connecting through commonalities; [it's] a lot of emotion in a 1x1 square. It's crazy!

—Lanie[1]

The 1x1 square referred to by Lanie Heller above is the Instagram image. Lanie, a long-time Instagram user, is remarking not simply on an image, or a form of expression, but also on a point of network, a point of affective exchange. Similarly, "photography must be understood simultaneously as a social practice, a networked technology, a material object and an image" (Larsen & Sandbye, 2014, p. xxiii). Starting from this manifold definition of photography and the multiplicity of the Instagram image, this chapter unpacks and analyzes hashtag publics on Instagram. Since it launched on 6 October 2010, Instagram, a Facebook-owned IB-MSN (image-based mobile social network)[2] centered on personal user–uploaded photographic content, has provided a framework for a heterogeneous mobile community out of a larger mobile public (Goggin, 2008). In December 2010, the ability to post searchable hashtags to photos was added, and the potential for organizing via hashtags came to fruition. This chapter argues that hashtags play a vital role in the production of intimate publics and are co-constitutive in the development and maintenance of communicative exchange between women on Instagram. It also demonstrates the ways in which a community of women shares intimate experiences through a platform that serves to marginalize women's experiences through

Figure 17.1. @_chemiko_ This is my five string serenade, 30 September 2014. #shootermag.

Figure 17.2. Sonja. We all go a little mad sometimes. Haven't you? 6 March 2014. #jj_bodybeautiful #fineartstorage #my_365 #crazycoolselfie #ma_portrait #ig_aau_fa132 #ig_gods #mextagram_selfie_plus #edits_sp #hitchcock.

Figure 17.3. Magdalena Olszanowski. @raisecain The lovely @momma2maxh featured me on @mobileartistry today! I am so honored. / I started my self-imagining practice online in the mid 90s as a teenager. I was fascinated by the way women like Francesca Woodman, Hannah Wilke and Carolee Schneemann were using their bodies as a way to interrogate the ideological conventions of femininity, gender relations & sexuality […] This photo is part of a feminist (mobile) conceptual series about the pleasure & trauma of doing my PhD called #beingPhD. If you'd like go to @mobileartistry for more details on my process 19 April 2013. #beingPhD #mobileartistry.

Figure 17.4. Lanie Heller. @momma2maxh Happy Monday…the hand pose series continues, sorry to say. 30 September 2013. #bws_edits, #rsa_dark, #creativebw_color, #rsa_ladies #momma2maxh-selfie #365sps #bws_edits #mobileartistry#iphoneonly #ipmrocks #soulboo #stillsickaftertwoweeks.

Figure 17.5. Jaclyn Turner. @jaclyn_t jet lag + sleep deprivation + deadlines + panic attacks + unrealistic expectations coming from all over and within, manifesting itself into my eyes, filled with redness and pain. 1 March 2014 #selfactualization #selfportrait #vsco #vscocam #vscolovers #blender #3dmax #architecture #shootermag #shootermag_canada #sundaybluesedit #bluesaturday.

Digital photographs with digital post-production alterations. Courtesy of the authors.

inconsistent modes of censorship.[3] Consequently, these arguments illustrate how the Instagram image must be read as a networked image to fully understand its ontology.

METHOD

Since 2011, I have corresponded with dozens of Instagram users about their use and about the network at large. For this chapter, using semi-structured interviews (Oakley, 1988), I focus on the Instagram practices of myself and four other women: Jaclyn Turner, Lanie Heller, Sonja,[4] and Chemiko.[5] The women I have interviewed are English speaking, have varying ethnic backgrounds, and range in age from 30 to 45. Only Jaclyn and I have had professional photography practices before we started Instagramming; Lanie went to art school and focused on analog photography before digital cameras were available, and the two others did not consider photography as a practice at all before Instagram. Instagram is not another outlet for their image practices; it has become the main platform for it. In addition, I conduct a discourse analysis of the subjects' images and their concomitant paratext, specifically focused on hashtag use. In doing so, I recognize that as part of their Instagram practice and community maintenance, some of the women are continually removing their images and rewriting the paratext of other images—a form of participation in the intimate public.

THE INTIMATE PUBLIC

Why are Instagram images such potent points of networking, and why does Instagram continue to draw numerous users despite its draconian policies and monitoring?[6] I turn to Lauren Berlant (2008), Rob Horning (2014) and danah boyd and Alice Marwick's (2011) ideas on networked publics, "publics that are restructured by networked technologies" (boyd, 2010, p. 39). Networked technologies in this case constitute mobile devices, transmission technologies, and mobile apps. The specific public I'm focused on is one that is highly connected with their mobile devices to Instagram. For the purposes of my arguments, I rely on the concept of community as defined by the women I interview, including their claims about IB-MSN labour, a form of unwaged labour they perform to keep their communities active. This is foregrounded later in the chapter as I unpack their ambivalent thoughts on the reward of hashtagging versus the amount of labour involved. Rob Horning (2014) acerbically points out that "apps' networks masquerade as ersatz 'communities,' but such networks actually constitute a medium designed to let them seek out precisely the people they can exploit" (n.p.). Horning's identification of community is critical here, because the chief reason these women continue to post on Instagram is *community*. However, for them, pleasure in belonging trumps the evident unwaged digital labour involved in using Instagram (Lazzarato, 1996).

The way my participants frame community evokes what Lauren Berlant calls an intimate public. She notes that "what makes a public sphere intimate is an expectation that the consumers of its particular stuff *already* share a worldview and emotional knowledge that they have derived from a broadly common historical experience" (2008, p. viii).[7] The "intimate public" is a produced space within which participants can exchange ideas about their ways of being in the world and also discipline those ways of being within that shared worldview. This specific intimate public produces networked spaces in a contested space. Instagram's policies and the policing of images by users on one hand create a hostile environment for women's feminist art practices (see Figure 17.1) but, on the other, create the conditions for these women to negotiate the policies, manage their own spaces, and find others in the circumstance of risk (hooks, 2014). Community as conceptualized by my subjects consists of comfort in experimenting with risky images and a space within which to do so, simultaneously recognizing that their space is always at risk because of the platform's policies—the very policies that maintain

Instagram's capital (Horning, 2014). Risky, in this case, means images that are nude, nonconforming to ideals of beauty, dealing with violent subject matter, and generally containing themes and aesthetics that could be flagged and/or removed by Instagram.

How are the conditions for safety managed? Instagram user Chemiko tells me: "It's the community you surround yourself with and immerse yourself into that can elicit this feeling of security." Particular types of communities are read as spaces that remove a "fear of being rejected, fear of being mocked or unfairly criticized."[8] In other words, as bell hooks (2014) argues, spaces need to produce comfort in the face of risk so women can share their work and "get on stage" without being shamed.[9] For these women, there is always a risk because the community always hinges on an Other that can enter the space, shame, and report images to Instagram. Chemiko is currently using her fourth Instagram account: her three preceding ones have been shut down by the platform. Her reticence to continue to use hashtags arises from the networked knowledge that (a) the digital labour[10] put in to archive, organize, and annotate images is time consuming,[11] (b) all of that work and all of those images can be removed in an instant without warning and without any way to retrieve the data.

Despite these risks, users still "get on stage" (see Figure 17.1). There is an attachment to Instagram. Berlant's theorization of intimacy combined with Horning's ideas of networked space helps provide a more comprehensive analysis of that attachment—"intimacy builds worlds; it creates spaces and usurps places meant for other kinds of relation" (Berlant, 1998, p. 282), such as places structured towards the benefit of capitalist business practices (Horning, 2014; Terranova, 2000). Intimacy organizes the conditions for risk taking and answers the question of why so many women continue to participate despite the platform's prominent censorship. Analogous to the women's understanding of the precarity of the platform, however, as demonstrated by Chemiko's reticence towards employing hashtags, "intimacy's potential failure to stabilize closeness always haunts its persistent activity, making the very attachments deemed to buttress 'a life' seem in a state of constant if latent vulnerability" (Berlant, 2008, p. 282). Hashtags facilitate the reading of vulnerability imbued in the images (see Figures 17.1–5). Hashtags assist both the viewer in guiding them to read the image in a particular way and the creator in providing them with a language to better understand their own image and its role as part of different communities. The image is never encountered solely as an image; it is encountered as a node in a network. The idea of community as theorized by my subjects can be read *in* the image: both the community that has been written into the structure of Instagram and the one the women activate and sustain. The networked image (the point of affective exchange) is imbued with layers of meaning that change over time and that operate for the purpose of connectivity.[12] The hashtag is one of the layers of meaning making within this milieu.

Intimate publics engaged in feminist exchange emerge as a result of photography and are maintained through both the image and the hashtag. Van Dijck (2007, p. 63) argues that some women (like the ones in my research) participate in "communal photographic exchanges" to mark their identity and engage in a reciprocity of belonging to a specific culture, in this case, the Instagram counterpublic (Fraser, 1992), leading to enculturation. That is to say, they have a comprehension of the markers and aesthetics to fit into specific hashtag groups. Sonja (Figure 17.2) says:

> I hashtag my images to place them where others with similar taste will see them, often to be featured, and, yes, ultimately to interact with these people. I stopped tagging [certain communities] because I felt my images no longer fit there.[13]

By "fit," Sonja means the image doesn't fit the aesthetic of the other users posting to that hashtag. The hashtag and a caption, she points out, is as important as the image itself for others and for herself to have a better understanding of the image. The hashtag is part of what makes the Instagram image a network image. As mentioned before, it also acts as signifier and as cultural context builder, producing the conditions of enculturation.[14] Hashtags allow the users to participate in a culture with others and also shape their own images (e.g., posting an image based on a hashtag theme such as #sundaybluesedit). The images these women create are part of the "cultural condition of perpetual modification" (Van Dijck, 2007, p. 68). One of the ways that modification occurs is at the level of the hashtag, which can substantially change the way an image is read. The women often work in a conceptual register such that the rhetoric of the image necessitates hashtags to "explain the image more," often using hashtags that they have become privy to through exchanges with other women.[15] Consequently, via these ways, the image is effective in being a networked image and participating in the intimate public.

WHY USE HASHTAGS?

There are many motivations for using specific hashtags. I have identified these four categories:

(1) Finding "like-minded people." This first category is the principal motivation for all my subjects.[16] Almost half of my sample does not share their work with people in their day-to-day lives. The hashtag can facilitate a conversation about the image posted beyond the echo chamber of a user's current followers.

(2) Finding inspiration in the hashtag feed. Other images tagged with a chosen hashtag make users consider their own practice and look for specific aesthetic arrangements that they would not otherwise. One of my

subjects dwells on this motivation, leading me to wonder how significant hashtag communities are to creating aesthetic trends.

(3) Participating in hashtag-based challenges, contests, and/or communities.
(4) Archiving collections of images.

Arguably, these four general categories produce and maintain Instagram. In the next section, I will explain the role of hashtags for the development and maintenance of feminist exchange by way of these four categories.

WOMEN FIND EACH OTHER

There are two direct ways to search Instagram: by hashtag and by user. All the women in my sample mention using hashtags to find other like-minded people and communities as the most important function of Instagram, particularly when they first joined. As users I follow in my feed post images with hashtags, I have the ability to click on those hashtags and see other users' photos. Using hashtags for my own photos, I am prompted to peruse each hashtag feed to observe (be inspired) and subsequently analyze the feed to note how my image fits into that specific milieu. Since several of the subjects are early adopters, their hashtag use has changed over time. In the early days of Instagram, Jaclyn used hashtags because the IB-MSN wasn't as widespread, and "now there is so much clutter [in the hashtag feed], I rarely bother to check." Yet she still maintains hashtags for nearly all the images she posts and participates in hashtag-based weekly challenges. She echoes the sentiment of the others,

> I definitely felt it was easier to reach people in the beginning…at least via hashtag. I feel like it's already crossed the point for that now…hashtagging now is mostly for archival purposes, and to explain the process[17] or for weekly challenges…like the colors of the week thing is one of my fav[orite] ways to use the hashtag now.[18]

All of the subjects I interviewed told me that hashtags were a useful way to find other users when they first started Instagram. Lanie recounts:

> I found a group of women who were suppressing their artistic desires because of children, but because of the iPhone it was easy to have everything in our hands. We discovered we could capture things and create quickly rather than with a darkroom, paint or sculpture. So we had this common bond. I'm still friends with a lot of the people I met in 2011.[19]

Lanie's excitement about the early Instagram hashtags and her first hashtag group, the feminist **#igsistas**, grew out of her finding other mothers through hashtags. Although Instagram's platform constitutes the conditions for communities to form, there are no preformed communities of this kind. These communities are imminent to the ways in which women take up using the network via their hashtag use.

Finding like-minded people also leads to acquiring new likes and/or follow-ers. This engagement works with the ways in which new social media platforms emphasize what Alice Marwick (2010) and Theresa Senft (2008) call the culture of self-branding and micro-celebrity. Sonja tells me: "You find one community leads to another and then others find you."[20] Other users suggest hashtags for fea-tures, and so on. Hashtag communities also inspire the act of taking photos. Sonja and Turner both use **#sundaybluesedit** consistently. They create images to fit with the theme of the hashtag (see Figure 17.5). Either they edit previous photos accordingly or take a new photo. Over time, users learn what fits a specific com-munity and how they should position their aesthetic (self-branding) to participate. If you don't tag, you can miss out on interaction. However, Sonja and Chemiko point out that by hashtagging, there's always the chance that the "wrong"[21] users can find you. When I first started writing about Instagram in 2012, I wrote about the ephemerality of hashtag communities: that if no one "upkeeps" the hashtag by continuing to post, it will stop being a community (Olszanowski, 2012). This is why many user-generated hashtags have moderators, and all of the women I inter-viewed are or have been hashtag community moderators. One of the activities of a hashtag moderator is to seek material for "exhibitions," called "features," in which they repost different users and their work on the hashtag's main account feed. For example, **#mobileartistry** has the account @mobileartistry that shares a curated selection of **#mobileartistry** tagged posts. This is done for a wider exposure of the user and their work; the way to be found or discovered is to hashtag your photo appropriately.[22]

While tagging a post with a popular hashtag increases its exposure, something all interviewees value, it also increases the chances of having an account flagged/removed and of attracting incidences of unwelcome comments/engagements and new, less engaged followers. In short, with hashtags, one can be made searchable, but that also means becoming a subject of surveillance (Page, 2012, p. 182).

Despite its community-generating benefits, the copious unwaged labour in moderating these communities has led some of the subjects to stop because, as Sonja explains from her perspective, "moderating is much effort and I'd prefer to spend more time on my own account creating images." She tells me that "when you're moderating it means sifting through [hashtags] to find people's submis-sions."[23] This can take hours and requires a comprehensive understanding of what fits into the community, what can bring about new connections, and so on. Arguably, the users are also engaging in a form of surveillance, like that men-tioned earlier in the chapter. They are setting standards and disciplining ways of use despite well-intentioned motivations. This creative digital labour also turns into emotional labour through messaging users to submit their work, explaining the community, creating relationships, and so on (Fuchs & Sevignani, 2013; Senft, 2008). For example, through moderating, Sonja noticed that users often have more than one account. She suggested that some people want to have alternate galleries to have those images separate and unidentifiable and to not alienate their current audience—a potential tactic to circumvent censorship (Olszanowski, 2014). If, as Lanie (see Figure 17.4) recounts, "[the 1x1 photo] reveals more about me than I would in everyday life,"[24] then it makes sense that some women would want to create alternate accounts depending on their followers and participate in ways that do not cleave to the normative paradigm of Instagram use. And if the women suggest that the paratext is as important, then we can suggest that hashtags contribute to identity formation and active subjectivity online (Page, 2012).

For several of the users, hashtags have also become archival repositories, a personal public space that contributes to the formation of their selves online.[25] Jaclyn recounts a hashtag shared with her partner, Leo, that one day started being used by another woman who knew Leo.[26] The other woman started posting photos of herself and her partner to the hashtag. The hashtag, which was focused on Jaclyn and Leo, had about 300 photos that belonged to them already. Jaclyn was frustrated at first and felt that the other woman took over her space until she started a new similar hashtag and accepted that with the scale of Instagram, it is futile to assume personal ownership of hashtags. It is hard to avoid other people using a hashtag that one has become accustomed and habituated to. By the same token, she still invests time in several personal hashtags. Lanie expressed surprise that some of her initial hashtags still *belong* to her and is hopeful that some of her personal hashtags will not be taken over by others.

HOW DO WE READ THE NETWORKED IMAGE?

As outlined above, all the women mention what Barthes (1977) calls a shared rhetoric of the image. This is built and learned through the semiotic landscape of language, form, aesthetic, and paratext. Barthes focuses on the advertising image. I read the Instagram image operating in a similar way to advertising; in the age of self-branding, the Instagram image is calling the viewer into the user's product. We can see the similarity in that the Instagram images often include all or some of the rhetoric identified by Barthes: the image (product), hashtags (which advertising now has started to adopt), copy (a description, a poem, a story), a heading (a pithy quote or aphorism), and/or links for further information.

Communities on Instagram are formed around the rhetoric of the image, which is most often underpinned by a hashtag or a series of hashtags. Without the hashtag, the image would not operate in the same way. The hashtag is the anchor for the networked image, that which tethers it to particular meanings and sustains a commonality across images. The anchor allows women to read images in particular ways. Jaclyn echoes (see Figure 17.5): "I find a lot of people I connect with share a lot of themselves."[27] How can Jaclyn know the other users are sharing "a lot"? What does sharing a lot entail? Chemiko tells me she hates small talk and has difficulty making friends. On Instagram, she is able to communicate through images, exhibiting the operation of the rhetoric of the image: "I see it as if [the other users] are revealing themselves in a similar manner (e.g., with specific types of poses, partially nude, etc.). Then chances are they would view mine in the same realm…it was easier for me to open up in this platform without feeling too uncomfortable or too shy."[28] Chemiko's explanation of the commonalities between user experiences foregrounds her understanding of networked life subjectivity. Similar to Horning (2014), Rainie and Wellman (2012) argue that "networked individuals are voluntarily creating content every day in tandem with other networked individuals within and outside of their own personal networks" (p. 200). They continue in a more generous stream and argue that this interaction "can expand and enrich collective knowledge" (p. 201), similar to the way Sonia Livingstone (2005) describes a public, one that is able to reach common understandings of shared interests. Based on my interviews, it is evident that users are finding hashtag communities to participate in and with, detached from the larger Instagram public. By doing so, these users are able to forge community and support systems for their work.

I first met Lanie when she wanted to feature one of my images on **#mobileartistry** (see Figure 17.3). We started chatting on Kik, a messenger service, and lost contact until I produced a photo I wanted to share with the **#mobileartistry** community. The way I read Lanie's images (see Figure 17.4) resonated with my worldview. I found myself in a state of vulnerability the intimate public enacts (Berlant, 2008) by sharing intimate details of my life with her and taking the risk to post more personal images. The Instagram image and its users operate in a paradoxical mode; they are intimate and personal while simultaneously public "by virtue of being shared on a social network" (Champion, 2012, p. 85). Lanie illustrates: "I wanted my name and images to stay in the app. I wanted to stay in that community, that was *my* place to unwind and get away. You don't want everyone else to search and see it."[29] Even though it is out in public, users still can feel as if their images and hashtags are meant for only certain users. Sonja, similarly to Jaclyn and Lanie, is quick to astutely point out, "the reality is, [the hashtag] is anyone's."[30]

The effort in keeping and participating in communities on Instagram, a form of unwaged labour, is a common issue among my subjects. Chemiko talks about the rewards of hashtags, suggesting there aren't any for her anymore and that it takes too much effort. Yet she does use hashtags to seek out others, a behaviour she finds rewarding. Seeking out others by participating in local hashtags is another way to participate in a community. **#igerskansas**, a hashtag Jaclyn started, is one she commonly peruses and posts to before, during, and after she travels.

She often meets users in other cities through their city-specific hashtags or from other hashtag communities. "Nearly all the friends I visit/stay with in NYC now are from IG, I never even call my university friends who live there," she laughs.[31] While Champion (2012) argues that this platform "effectively disallows Instagram photographs the function of documenting past events" because of the associated lack of a social relationship between the photograph, photographer, and viewer(s) (p. 86), my analysis and observations demonstrate that this is not necessarily the case. Hashtag communities on Instagram are formed through their pasts. Hashtags allow a user to explore an archive to peruse tagged photos and/or to participate in the process of enculturation. This networked life is seen in the subject's images and the associated paratext.

For example, Lanie's black-and-white self-portrait (Figure 17.4) has 33 comments, 17 from other users and 16 from Lanie, which shows a strong engagement with her followers. This is typical of all the women I interviewed. For instance, Sonja apologized for unfollowing me but informed me I was not posting and we no longer interacted regularly. I was posting once every week, and that lag in Instagram time is too much for some users. The labour involved in keeping up your account, posting, and interacting with people is immense. Sonja informed me that she unfollows people who are not regularly engaging with her. What constitutes regular engagement is unclear, as none of the subjects was able to give me a time frame, but echoed that it is a feeling of "disconnection" from the other user. Terranova (2000) argues that the Internet is "about the extraction of value from updateable work" which is labour intensive (p. 48).[32] It is not enough to have a website or a blog or, in this case, an Instagram account; one must be continually updating it with new content to keep up with the aesthetic and economy of the new to (1) not lose one's audience and (2) gain new audience members. Demand for fresh content from other users means a continuous stream of new images. Does the rhetoric of community become almost too much for the very community members it recruits?

THE CHOREOGRAPHY OF THE NETWORKED IMAGE

Annette Markham (2013), when discussing digital life, argues that to "recognize our own existence in any meaningful way, we must be responded to." A networked image's fruitfulness is in its ability to be responded to. It is unsurprising, then, that Sonja's feed of 179 posts has only one post with no hashtags, and it is because she @s all the hashtag communities in the image's description. Theorizing the Instagram image as a network image foregrounds its circulatory ontology. An image's circulation within the intimate public I have outlined allows us to read the Instagram image "simultaneously as a social practice, a networked technology, a material object and an image" (Larsen & Sandbye, 2014, p. xxiii). This chapter emphasizes the choreography of the IB-MSN ecology: how the media is used, what it does, and how it encourages women to take up modes of expression they did not think were available to them previously. Women are exposed to these modes through hashtags, which allow them to participate in a larger mobile public. As these intimate publics flourish on privately owned networks, we must not overlook the political economy of the communicative exchange. The communicative exchange that undergirds the Instagram image, which must be read as a point of network and a point of affective exchange, is regulated by the app's policies. However, the very policies and regulations that Instagram enacts also create the conditions possible to form communities strengthened by the very nature of that censorship, which is in part due to the hashtag.

NOTES

1. Lanie Heller, interview with author, 2014.
2. I derive my term "IB-MSN" from MSN. See de Souza e Silva and Frith (2010).
3. See Olszanowski (2014) for an analysis of Instagram's censorship policies.
4. Name has been changed to protect anonymity.
5. Instagram user name used to protect anonymity.
6. See Hestres (2013) for the tangled ways in which policies are set and enacted by the iOS store.
7. The community was built through these women's practices. The community, as Berlant delineates, is also a space that exists because of the consumer market. Without the platform, it is uncertain if and/or how these women would engage in photographic practices.
8. Chemiko, interview with author, 2014.
9. This is hooks's answer to the notion of "safe space," which she argues is a concept that assumes *a priori* conditions for the production of safety and also belongs to the privilege of White feminism.
10. On digital labour, see Terranova (2000).
11. Sonja also explains how tiring keeping up with hashtag communities has become for her.
12. For example, Lanie met her husband on Instagram through the **#abnormal** hashtag.
13. Sonja, interview with author, 2014.
14. For instance, http://iconosquare.com/ provides detailed user data and metrics. Under their "optimization" menu, a subheading of "Tag Impact" reveals your top tags and the most frequently used tags on Instagram. The user's top tags that correspond with Instagram's top tags are highlighted. These metrics are part of what Senft (2008) calls micro-celebrity culture.
15. Sonja, interview with author, 2014; Jaclyn, interview with author, 2014.
16. All interview subjects used the term "like-minded people."
17. Such as specifying what camera, app, and/or filter is used. Jaclyn tags a lot of her images with **#vscocam**, a popular app for editing photos.
18. Jaclyn Turner, interview with author, 2014.
19. Lanie Heller, interview with author, 2014.
20. Sonja, interview with author, 2014.

21. Users that will flag your photo or troll your feed.
22. This is how I met Lanie. I was following the account @mobileartistry and its hashtag #mobileartistry, and one day she commented under one of my images, asking if I would like to be featured.
23. Sonja, interview with author, 2014.
24. Lanie Heller, interview with author, 2014.
25. For a robust discussion of self-representation online, see Walker Rettberg (2014).
26. Jaclyn met Leo through a local hashtag on Twitter before Instagram started. His name has been changed.
27. Jaclyn Turner, interview with author, 2014.
28. Chemiko, interview with author, 2014.
29. Lanie Heller, interview with author, 2014.
30. Sonja, interview with author, 2014.
31. Jaclyn Turner, interview with author, 2014.
32. Sonja laments this point: "I learned so much [from those hashtag communities] and began experimenting more and more. It was tiring....All required the main hashtag plus the daily tag." By that, she means that in order to participate, her account had to keep up with all the different hashtags being used and promoted by that community. If you were part of several communities, you could spend hours taking, editing, and circulating photos that fit the hashtag theme.

REFERENCES

Barthes, R. (1977). *Image, music, text* (S. Heath, Trans.). New York: Hill & Wang.

Berlant, L. G. (1998). Intimacy: A special issue. *Critical Inquiry, 24*(2), 281–288.

Berlant, L. G. (2008). *The female complaint: The unfinished business of sentimentality in American culture.* Durham, NC: Duke University Press.

boyd, d. (2010). Social network sites as networked publics: Affordances, dynamics, and implications. In Z. Papacharissi (Ed.), *Networked self: Identity, community, and culture on social network sites* (pp. 39–58). London: Routledge.

boyd, d., & Marwick, A. (2011). *Social privacy in networked publics: Teens' attitudes, practices, and strategies.* Paper presented at the 4th Annual Privacy Law Scholars Conference, June 2–3, Berkeley, CA.

Champion, C. (2012). Instagram: je-suis-là? *Philosophy of Photography, 3*(1), 83–88.

de Souza e Silva, A., & Frith, J. (2010). Locative mobile social networks: Mapping communication and location in urban spaces. *Mobilities, 5*(4), 485–505.

Fraser, N. (1992). Rethinking the public sphere: A contribution to the critique of actually existing democracy. In C. Calhoun (Ed.), *Habermas and the public sphere* (pp. 109–142). Cambridge, MA: MIT Press.

Fuchs, C., & Sevignani, S. (2013). What is digital labour? What is digital work? What's their difference? And why do these questions matter for understanding social media? *tripleC: Communication, Capitalism & Critique, 11*(2), 237–293.

Goggin, G. (2008). The models and politics of mobile media. *The Fibreculture Journal, 12.* Retrieved from http://twelve.fibreculturejournal.org/fcj-082-the-models-and-politics-of-mobile-media/

Hestres, L. (2013). App neutrality: Apple's App Store and freedom of expression online. *International Journal of Communication, 7,* 1265–1280.

hooks, b. (2014, October 8). *A public dialogue between bell hooks and Laverne Cox*. Retrieved from New School website: http://new.livestream.com/TheNewSchool/bell-hooks-Laverne-Cox

Horning, R. (2014, June 20). "Sharing" economy and self-exploitation. *New Inquiry*. Retrieved from http://thenewinquiry.com/blogs/marginal-utility/sharing-economy-and-self-exploitation/

Larsen, J., & Sandbye, M. (2014). The new face of snapshot photography. In J. Larsen & M. Sandbye (Eds.), *Digital snaps: The new face of photography* (pp. xv–xxxii). London: IB Tauris.

Lazzarato, M. (1996). Immaterial labor. In M. Hardt & P. Virno (Eds.), *Radical thought in Italy: A potential politics* (pp. 133–147). Minneapolis: University of Minnesota Press.

Livingstone, S. (2005). *Audiences and publics: When cultural engagement matters for the public sphere*. Portland, OR: Intellect.

Markham, A. (2013). Dramaturgy of digital experience. In C. Edgley (Ed.), *The drama of social life: A dramaturgical handbook* (pp. 279–94). Farnham, Surrey, UK: Ashgate Press.

Marwick, A. (2010). Status update: Celebrity, publicity and self-branding in Web 2.0. *Dissertation Abstracts International*, 71–12.

Oakley, A. (1988). Interviewing women: A contradiction in terms. In H. Roberts (Ed.), *Doing feminist research* (pp. 30–61). London: Routledge.

Olszanowski, M. (2012). *Instagram's 'Space of Flows'*. Paper presented at the 13th Annual Conference for the Association of Internet Researchers, October 18–21, Manchester, UK.

Olszanowski, M. (2014). Feminist self-imaging and Instagram: Tactics of circumventing sensorship. *Visual Communication Quarterly*, *21*(2), 83–95.

Page, R. (2012). The linguistics of self-branding and micro-celebrity in Twitter: The role of hashtags. *Discourse & Communication*, *6*(2), 181–201.

Rainie, L., & Wellman, B. (2012). *Networked: The new social operating system*. Cambridge, MA: MIT Press.

Senft, T. (2008). *Camgirls: Celebrity and community in the age of social networks*. New York: Peter Lang.

Terranova, T. (2000). Free labor: Producing culture for the digital economy. *Social Text*, *18*(2), 33–58.

Van Dijck, J. (2007). *Mediated memories in the digital age*. Stanford, CA: Stanford University Press.

Walker Rettberg, J. (2014). *Seeing ourselves through technology: How we use selfies, blogs and wearable devices to see and shape ourselves*. New York: Palgrave Macmillan.

CHAPTER EIGHTEEN

Meta-Hashtag and Tag Co-occurrence: From Organization to Politics in the French Canadian Twittersphere

SYLVAIN ROCHELEAU AND MÉLANIE MILLETTE

INTRODUCTION

French and English are the two official languages in Canada. Even though the majority of Canadians are English speaking, French Canadians account for more than 7 million of the total population (21.3%). The largest Francophone community is concentrated in Quebec (6 million), the only province where the only official language is French. The remaining French Canadians live outside Quebec, scattered throughout English-speaking provinces and territories (Statistics Canada, 2014). They have to deal with limited resources to keep their community networks alive, defend their culture, and campaign for the preservation of linguistic rights. Social media represent a nonexpensive, accessible opportunity to pursue these activities, as many French community–based webzines, blogs, and Twitter and Facebook accounts demonstrate. Here, we focus on the use of the microblogging platform Twitter by French-speaking Canadian minority populations.

Drawing from data harvested from Twitter and from interviews, this study investigates how these minority populations use hashtags to organize their online communication. We crafted a long-term methodological strategy, and over time we observed that many different hashtags were used in the community, often simultaneously in one message. On Twitter, messages cannot exceed the maximum length of 140 characters, and using many hashtags "consumes" characters. Why use so many hashtags in a single message? What is the significance of hashtag "co-occurrence" in an online political and linguistic community such as the French Canadian one? In this paper, we argue that hashtag co-occurrence can be understood as an organizational strategy, allowing users in a hashtag community to manage information while broadcasting it. We highlight how one specific hashtag, **#frcan,** is used as a "meta-hashtag" to identify a main theme on Twitter related to political French Canadian discussions and comments. This allows us to explore the collective aspect of hashtags.

We start by exploring the literature on Twitter's specific nomenclature and hashtag uses, from organizational purposes to collective subjectivity. We then present our methodology, taking advantage of a mainly qualitative method with a data-mining component. From there, we analyze the "meta" organizational purpose of the **#frcan** hashtag and its relation to other tags to explore the significance of tag co-occurrence. We propose a hashtag co-occurrence typology, organized under five main aspects corresponding to a tag's role in the observed online communication. We end this paper with a discussion about community and hashtag publics.

TWITTER'S POLITICAL USES

Twitter Nomenclature and the Importance of Hashtags

Online, tags are usually selected a posteriori to describe content, facilitate retrieval of the content, and archive it—using Delicious, EverNote, or other online clipping and retrieving tools. "In contrast, tagging has emerged as a method for filtering and promoting content in Twitter rather than as a tool for recall" (Huang, Thornton, & Efthimiadis, 2010, p. 173). On Twitter, tagging happens while writing a tweet, not after broadcasting it. Taking advantage of this "tagging while broadcasting" state of affairs, Starbird and Stamberger (2010) proposed a Twitter-based syntax to "tweak" tweets for crisis-reporting efficiency. Without specifically using the term "co-occurrence," these researchers proposed multi-tagging as a way to organize information and optimize online citizen reporting during crises. The multiple hashtags in a tweet allow the users to structure and manage their information, supporting uncoordinated civic practices.

Political Hashtags

The political use of hashtags in Twitter has been explored in social sciences and is still an object of interest in current studies. The recent movements in Middle

East gave an opportunity to observe how a large number of uncoordinated users can create a collective narrative around a specific hashtag such as **#Egypt** (e.g., Papacharissi & Oliveria, 2012). Social crisis, protest, large-scale events and ecological disasters are other contexts where citizens use hashtags to build and organize a political discourse over a short period of time (e.g., see Vicari, 2013).

Analyzing tweets collected over a 4-month period, Small (2011) has shown how Canadians use the hashtag **#cdnpoli** (the most popular—and mainly English-driven—tag about politics in Canada) as an aggregator. She also noticed that contributors tend to use many hashtags in a single tweet in order to "seek to get their message out to as many people as possible" (p. 884).

Hemsley and colleagues (2012) worked on the **#Occupy** Twitter stream to see whether there was any organizational pattern in tag co-occurrence during the Occupy Wall Street movement. Drawing on data collected over 20 days, they found that tag co-occurrence in Twitter is important in terms of location and relation to utility (e.g., controlling for distance). This study is one of the few academic works exploring specifically tag co-occurrence on Twitter. So far, tag co-occurrence—a quite common practice in Twitter— has received little attention from academia.

Hashtag Communities and Ad Hoc Publics

As Small (2011) has mentioned, users are aware that hashtags are a way to broadcast information to other citizens. The organizational purpose of hashtags can also come with a more subjective, social aspect. In this sense, users can share a form of collective consciousness around a specific hashtag, described as "ambient affiliation" by Zappavigna (2011). Her linguistic approach to Twitter demonstrates that the shared use of a hashtag "presupposes a virtual community of interested listeners who are actively following this keyword or may use it as a search term" (p. 791). In other words, hashtag use can imply a sense of collectiveness, a sense of an audience who, even if its contours are blurred, shares the same interest.

This aspect connects to the debate about the way we should describe Twitter users clusters. Here, we draw from Bruns and Burgess (2011),[1] who describe hashtag-organized parts of the political Australian Twittersphere as "topical hashtag communities" and "ad hoc publics." As they explain, "The term 'community'…would imply that hashtag participants share specific interests, are aware of, and are deliberately engaging with one another" (p. 5). Their analysis suggests that hashtag communities play a meso-level role in Twitter communication: they are permeable spaces, overlapping with wider networks, "macro-level flow of messages" articulated around long-term social bonds, and with the smaller, micro-level of human interactions enacted in @replies between people who don't know each other and never would have found each other without the hashtag (p. 6). Ad hoc publics, which may or may not be a subset of a hashtag community or even exceed it, are a space where users comment about a common interest without knowing each other or addressing each other.[2] These ad hoc publics of active users are characterized by a rapid formation around specific news or events (identified by "topical hashtags"), and they fade quickly (p. 2).

When it comes to a wider group of users, such as those using **#frcan** or **#cdnpoli**, it's hard to state if they are a community or a public. Some of them know others and interact with them; others just use the tags to disseminate information without interacting. A hashtag community behind **#cdnpoli** may contain (and overlap with) many ad hoc publics commenting on a bill project, an election, etc. Even if the social form changes, this illustrates that the hashtag is a significant nomenclature element when it comes to discussing politics online.[3]

Building on Small (2011), Hemsley et al. (2012), and Bruns and Burgess (2011), we propose that tag co-occurrence under a general meta-hashtag can be understood as an effective strategy for allowing a hashtag community—here, the French Canadian political sphere—to organize its communication into different flows or subthemes, such as provincial politics and to take advantage of emerging ad hoc publics.

METHODOLOGY: A MAINLY QUALITATIVE APPROACH TO FRAMING DATA MINING

Long-Term Observation-Participation and Preliminary Interviews

Our challenge was to consider both hashtag use and subjective aspects of the French Canadian Twittersphere. As a minority in terms of both demography and privilege, the French Canadian online population doesn't generate thousands of tweets per day. To circumvent this aspect, we built a mixed-method, mainly qualitative study and decided to collect data over a long period of time in order to generate meaningful findings. Our mixed method includes long-term online observation, preliminary interviews with members of the community, and, finally, data mining from the Twitter API.

One of the researchers observed French Canadian users on Twitter from December 2011 to May 2014. A Twitter user since 2007, she started in December 2011 by identifying 12 accounts from French Canadian NGOs, provincial institutions' accounts, grassroots accounts, etc., and aggregated them in a private Twitter list. She observed the tweets on a weekly basis, at least twice a week, during the first year. Over the first 10 months of the observation, this list grew to 102 Twitter accounts. The users behind these accounts were very heterogeneous—much like the online French-speaking minorities in Canada. The majority of the accounts followed were from grassroots groups, such as citizens' groups attached to a specific city, university, or school. The remainder was composed of official associations, French minority media accounts, and journalists, activists, or ordinary citizens interested in the "French fact" in Canada, and even some radical linguistic activists.

From an observant-participant position (Soulé, 2007), the researcher read the tweets broadcasted by these accounts for another year and a half. She sometimes interacted with users, and she collected the tags used by the community. The most frequently used hashtags and keywords and the most active accounts were identified as the starting point for the Twitter data mining. To validate these entry points, the researcher ran preliminary interviews with participants in the French Canadian community. The five interviewees were selected based on their strong relationships with different French-speaking areas. These exploratory interviews helped validate the relevance of the accounts, the hashtags, and the keywords selected to become the basis for our original query to the Twitter-streaming API. This list grew over the test period and was enriched by other keywords, accounts, and hashtags to 14 different items for the query.

From Data Mining to Meta-Hashtags

From January 21 to May 17, 2013, every tweet containing one or more hashtags or words from the list was saved in a relational database, for a total of 8,316 tweets. Tweets come with metadata such as username, location, date and time, and language used. All these metadata were classified in specific tables that we used to analyze and generate reports. To adequately assess the use of hashtags and co-occurrence, we also parsed every tweet to extract the hashtags and saved them in a particular table in order to search for patterns in the distribution of hashtags for this particular community.

The most popular tag from our sample was **#OnFr** (*Ontario français*), occurring 2,224 times. **#Frcan** was the second most popular hashtag with 2,039 occurrences.

In Canada, the province of Ontario is the largest, with close to a third of the country's total population (Statistics Canada, 2014). Due to its proximity with the Francophone province of Quebec, it also has the largest French-speaking population outside Quebec. This great concentration of Francophones in Ontario makes their regional hashtag, **#OnFr**, the most popular. Despite this, our long-term observation led us to identify **#frcan** as the meta-hashtag for its broader use to identify general communication regarding French Canada.

The predominance of a meta-hashtag under a large-scale theme in a given territory has been noticed in other research, such as Small with **#cdnpoli** (2011). Bruns and Highfield (2013) similarly observed that **#auspol** was an "umbrella hashtag" to broadly refer to the Australian political scene.[4] Thus, **#frcan** was a relevant starting point to observe tag co-occurrence and organizational capability in terms of geography and topics in our sample.

To take a closer look at this phenomenon, we filtered our data based on this meta-hashtag. From our original sample of 8,316 tweets, 2,039 tweets containing **#frcan** were filtered out to build our subset. Tweets from this subset were then parsed, and every time a hashtag was found, it was saved in a different table in our database in order to analyze the level of co-occurrence with the **#frcan** hashtag. We manually coded the most frequent tags co-occurring with **#frcan**.

DISCUSSION ABOUT CO-OCCURRENCE, META-HASHTAGS AND TAGGING AS AN ORGANIZATIONAL STRATEGY

Co-Occurrence Under #Frcan

Tweets containing the **#frcan** meta-hashtag showed a higher than average co-occurrence compared to a random sample of tweets: the average co-occurrence for the **#frcan** group was 2.68, compared to average co-occurrences 0.28 in a control group.[5]

Here, to properly understand what these results mean, we must acknowledge our sample bias. Since our three initial starting points to query the Twitter API were hashtags, keywords, and user accounts, and since hashtags are more numerous than the other two, our average co-occurrence result is probably higher than in the actual situation online. Nevertheless, the average tag co-occurrence we obtained is so much higher than the control sample that even if our results are slightly biased, they still demonstrate a tendency in the French Canadian political conversation. Co-occurrence was also noticed by Bruns and Burgess (2011, p. 5), when users add multiple tags in their messages "in order to ensure that it is visible to the largest possible audience."

One potential explanation that accounts for the high average of tag co-occurrences in the **#frcan** subsample is that multi-tagging helps to structure online conversations, therefore reducing the noise generated by the 500 million tweets broadcast each day. To explore this explanation, we manually coded the first 100 tags co-occurring with the **#frcan** meta-hashtag, based on the following categories:

Tag relative to ephemeral topics:
– Events and festivals
– News (e.g., **#budget2013**)
– Bill discussion (e.g., **#PL14**)

Tag relative to geography:
- Country (Canada or others)
- Province or territory of Canada
- City

Government-level or politics-related tags:
- Federal politics
- Provincial politics
- Federal parties
- Provincial parties
- Political long-term issues (*#emploi* [jobs])

Tag relative to cultural and linguistic identity:
- Cultural identity (e.g., *#FrancoOntarien* [French Ontarian])
- Identity aspects (e.g., **#ProudCanadian**)
- *Francophonie*
- Bilingualism
- Arts and culture
- Racism

Tag relative to civic life:
- Canadian institutions (e.g., *#BanqueDuCanada*)
- Media
- Education
- Emotion

Our interpretation of the tags and the selection of codes were informed by our long-time observation and the exploratory interviews conducted earlier. A tag could be coded with more than one category, such as **#eduYukon**, referring to both education and a geographic territory. We compiled these coded results and found that a quarter of the tags co-occurring with **#frcan** had a geographical meaning—either a province or territory name–derived tag such as **#frab** for French Alberta, or a form of location marker such as **#yyc**, referring to Calgary Airport international code—for a total of 1,170 tweets. Out of 100 tags, 22 different tags were political hashtags associated with levels of government, mainly federal (15 different tags, such as **#cdnpoli**) and provincial politics (6 different tags, such as **#NBpoli** and **#mbpoli**), for a total of 700 tweets. We had thought that event and news management would be more prevalent, but only 13 different tags, present in 200 tweets, were related to this organizational function (e.g., **#budget2013** and **#ottnews**). Surprisingly, tags related to civic life, including cultural and linguistic identity, ranked highest: 28 different tags out of 100 were related to the *Francophonie*, French Canadian culture, or bilingualism, for a total of 625 tweets in the subsample. Even though hashtags seem to be equal entities from the strict reading-level point of view, these results illustrate that they fulfill a complex organizational purpose.

Meta-Hashtag and Tags' Role: A Typology

The following table illustrates the role of the tags co-occurring with **#frcan** in our sample. Under this meta-hashtag, we found four other dominant roles, quite close to the tag's literal meaning:

Table 1. Hashtag Co-occurrence Typology.

Type of co-occurring tag	Role	Example from our data
Meta-hashtag	Main hashtag in a specific community or network. Used to differentiate the tweets relevant for this community from the entire Twitter stream.	#frcan
Civic and cultural markers	Tags used to characterize or to distinguish a subject or an object in a given online community. In our case, these cultural markers were mainly connected to the specificity of the French Canadian identity.	#francophonie, #franco, #bilinguisme
Location markers	Tags used to locate the tweets in a geographical context. These tags may or may not contain a political or cultural connotation (e.g., "fr" for French added to "on" for Ontario). They fulfill a role of geographical markers.	#canada, #OnFr, #ottcity, #Banff, #acadie, #ipefr, #frbc
Political markers	Tags used to identify the level of government the tweet refers to or their corresponding politics and political topics. They usually have a long-term presence in Twitter.	#canpol, #PolCan, #Ollo, #onpoli, #cdnpoli, #abgov, #mbpoli, #CRTC
Topical markers	Ephemeral tags associated, for the most part, with news, events and current affairs. Commonly used to promote an event or comment on breaking news.	#budget2013, #CauseCaron, #ottnews, #FinieLapathie

Users in our sample "tweak" their tweets by adding many tags, as Starbird and Stamberger (2010) suggested, to intentionally format their messages so other concerned citizens can find them. In this regard, our results confirm and enhance the findings of Hemsley et al. (2012) concerning the organizational potential of tag co-occurrence for location. Since "tags in Twitter are primarily used to find messages from other users about a topic" (Huang et al., 2010, p. 173), hashtag co-occurrence can be understood as a strategy for organization in an online political community.

Coming back to the specificity of our case study, we would like to formulate final comments about the collective subjectivity behind these tags.

From Community to Visibility: Toward Hashtags Politics

The French Canadian users may be a persistent hashtag community on Twitter, in the sense that **#frcan** acts as an aggregator for these users to find each other and comment about politics. Their intensive use of tags, as the high tag co-occurrence average in our sample demonstrates, is also significant in terms of collective subjectivity and visibility.

As Bruns and Burgess (2011) also noticed, tagging practices "underline the interpretation of using a thematic hashtag in one's tweet as an *explicit attempt to address an imagined community* of users who are following and discussing a specific topic" (p. 4; emphasis added). But our long-term observation and participation in the community leads us to conclude that this notion of imagined community should not be seen has a homogenous collective. Although they do share a main interest in French-speaking Canada, users we have interacted with represent a very heterogeneous group, with different political views and agendas—just as French-speaking minorities are. Thus, we have to consider the politics of hashtags, where hashtags are embedded in complex power relationships.

The linguistic aspect of our research highlights the complexity of the situation. As political power at the federal level is largely English based, some tags are used specifically to jump into an ongoing thread sustained by a wider English-speaking audience. English hashtags co-occurring with French ones often correspond to visibility tactics (Millette & Rocheleau, 2014). As an example, the tag **#polcan** (143 co-occurrences with **#frcan** in our subsample) is an attempt to connect the French political conversation to the general Canadian one—mainly English—and to bring French Canadian issues to the attention of the dominant Canadian Twittersphere.

Similarly, the tag **#francophonie** is the meta-hashtag of another wider community, namely those in French-speaking nations worldwide. Mentioning this hashtag in co-occurrence with **#frcan** potentially draws the attention of other Francophones to the French Canadian situation whilst adding the Canadian point of view to a wider conversation in the French-speaking Twittersphere. In these two cases, co-occurrence could be understood not only as an organizational strategy between different conversations from different communities using different hashtags but also as an act of political visibility.

CONCLUSION

Even though the French Canadian hashtag Twittersphere is small, it has developed a wide variety of hashtags. An analysis of the most popular hashtags in the community led us to build a subsample to specifically observe the tag co-occurrence structure under the dominant meta-hashtag **#frcan**. This meta-hashtag revealed a high average of hashtag co-occurrence, leading us to propose a tag typology based on five main different types of tags, depending on their roles: meta-hashtags, civic and cultural markers, location markers, political markers, and topical hashtags. This typology is useful to understand the role of tags as an organizational nomenclature on Twitter in a given online political community.

As French Canadian minority populations have a distinct and precarious status, especially outside the province of Quebec, their cultural and linguistic specificities become very emotionally and politically charged. This may explain why this code category, interpreted as civic and cultural markers, was the most present co-occurring marker in our **#frcan** sample. Our long-term and mainly qualitative approach was decisive in our capacity to grasp such a subtle organizational strategy. Future research should continue to demonstrate methodological long-term engagement to better understand hashtag politics.

NOTES

1. See also Chapter 1, this volume.
2. Although Bruns and Burgess imply that a hashtag community can turn into an ad hoc public when a breaking news hashtag emerges and is used by it, they don't illustrate how social interactions between users change during this process in order to justify the switch from a community to a public form. They built a useful comparison between issue publics and ad hoc publics, but this doesn't sustain a sociological understanding of a public as a political active agent, resulting from the subjective knowledge of a shared reading. This aspect should be considered theoretically in itself.
3. On the types of collectivities and networks that hashtags can enable and act within, see also Rambukkana, Introduction, this volume.
4. See also Sauter and Bruns, Chapter 3, this volume.
5. The other control sample harvested random tweets with the basic words "the" and "*le*."

REFERENCES

Bruns, A., & Burgess, J. E. (2011). *The use of Twitter hashtags in the formation of ad hoc publics*. Paper presented at the 6th European Consortium for Political Research General Conference, August 25–27, University of Iceland, Reykjavik.

Bruns, A., & Highfield, T. (2013). Political networks on Twitter: Tweeting the Queensland state election. *Information, Communication & Society, 16*(5), 667–691.

Hemsley, J., Thornton, K., Eckert, J., Mason, R. M., Walker, S., & Nahon, K. (2012). *30,000,000 Occupation Tweets: A hashtag co-occurrence network analysis of information flows*. Paper presented at IR 13.0, October 18–21, Manchester, UK.

Huang, J., Thornton, K. M., & Efthimiadis, E. N. (2010). Conversational tagging in twitter. In *Proceedings of the 21st ACM conference on Hypertext and hypermedia* (pp. 173–178). New York: ACM.

Millette, M., & Rocheleau, S. (2014). Tactiques de mise en visibilité : Usage de Twitter par des acteurs des minorités franco-canadiennes. In S. Zlitni, F. Liénard, D. Dula, & C. Crumière (Eds.), *Actes du Colloque international Communication* électronique, *Cultures et Identités (CECI)* (pp. 487–504). Le Havre, France: Éditions Klog.

Papacharissi, Z., & de Fatima Oliveira, M. (2012). Affective news and networked publics: The rhythms of news storytelling on #Egypt. *Journal of Communication, 62*(2), 266–282.

Small, T. A. (2011). What the hashtag?: A content analysis of Canadian politics on Twitter. *Information, Communication & Society, 14*(6), 872–895.

Soulé, B. (2007). Observation participante ou participation observante? Usages et justifications de la notion de participation observante en sciences sociales. *Recherches Qualitatives, 27*(1), 127–140.

Starbird, K., & Stamberger, J. (2010). Tweak the tweet: Leveraging microblogging proliferation with a prescriptive syntax to support citizen reporting. *Proceedings of the 7th International ISCRAM Conference*, Seattle, WA.

Statistics Canada. (2014). Population selon la langue maternelle et les groupes d'âge (total), chiffres de 2011, pour le Canada, les provinces et les territoires. Retrieved from http://www12.statcan.gc.ca/census-recensement/2011/dp-pd/hlt-fst/lang/Pages/highlight.cfm?TabID=1&Lang=F&Asc=1&PRCode=01&OrderBy=999&View=1&tableID=401&queryID=1&Age=1

Vicari, S. (2013). Public reasoning around social contention: A case study of Twitter use in the Italian mobilization for global change. *Current Sociology, 61*(4), 474–490.

Zappavigna, M. (2011). Ambient affiliation: A linguistic perspective on Twitter. *New Media & Society, 13*(5), 788–806.

CHAPTER NINETEEN

The Twitter Citizen: Problematizing Traditional Media Dominance in an Online Political Discussion

BRETT BERGIE AND JAIGRIS HODSON

INTRODUCTION

Wide-ranging public discourse among diverse actors makes an important contribution to the vitality of a democratic society. Castells (2008, p. 78) observed that "[w]ithout an effective civil society capable of structuring and channeling citizen debates over diverse ideas and conflicting interests, the state drifts away from its subjects." Since the popular adoption of Web 2.0 technologies or the participatory web, scholars have wondered about its promise and limitations in forming a participatory and heterogeneous democratic discussion space (Bohman, 2004; Castells, 2008; Drache, 2008; Pariser, 2011; Small, 2011). Importantly, these tools have the potential to transform people's relationships with media (Benkler, 2006; Jenkins, 2006) and form a venue for bidirectional discussions that bypass government and traditional media controls (Castells, 2008; Drache, 2008). However, questions remain. Is the promise of this technology fulfilled? Is it really bidirectional and open to all, or is the practice of participation on such platforms dominated by the same communication flows, as is seen in more traditional media sources such as television and newspapers?

Our study aims to answer questions about the nature of public discourse currently occurring on Twitter, a micro-blogging and social networking platform. Since 2006, Twitter has grown to host 255 million monthly active users who disseminate 500 million tweets per day (Twitter, 2014). Our study examined Twitter hashtag discussions relating to the 2013 budget presentation of the government of Alberta, Canada. A provincial budget touches all areas of public policy and spending and attracts interest and scrutiny from a multitude of powerful agents (e.g., political parties, traditional mass media outlets, labour unions, and advocacy groups) with reach across the public. Within this context, potential exists for a broad public to organize around relevant discussion, as delineated through hashtag use, on Twitter.

We collected 2,608 relevant tweets over a 3-day period that corresponded with the 2013 Alberta budget announcement and contained either or both the **#AbBudget** and **#Budget2013** hashtags, using Tweet Archivist (http://www.tweetarchivist.com/), an open-source software tool. Content and discourse analysis were employed to understand the content of the tweets and determine whether they represented activity congruent with civil society discussion. Our analysis suggests that Twitter may be an ill-suited discussion platform for civil society. Discussion uptake and opinion expression were relatively modest among participants. Furthermore, signs of inequality among discussion participants emerged: traditional mass media agents dominated the discussion with adverse affects, including permeating the discussion with traditional media content and fragmenting the community by driving participants to traditional media platforms in the service of economic interests. By consequence, the discussion that unfolded on Twitter was driven by the private interests of traditional media outlets, an outcome incongruent with the functional characteristics of the public sphere (Habermas, 2006).

TWITTER AND PUBLIC DISCOURSE

Social media allow users to form social networks based on interests and interpersonal relationships and allow them to exchange information, news, and other user-generated content (Kim, Hsu, & Gil de Zúñiga, 2013). The means of producing content and reaching large networked publics are widely available (Gerhards & Schäfer, 2010). Networked media thus can reorient individuals from passive consumers to critical observers and participants in a conversation (Benkler, 2006). The potential effect is that social media afford the type of communication that could shift power away from traditional mass media, through which owners enjoy inordinate influence on shaping public opinion (Benkler, 2006; Gerhards & Schäfer, 2010; Jenkins, 2006).

Twitter allows users to create "tweets" of up to 140 characters, which appear chronologically in the feeds of the other users that choose to follow their updates. Through cross-referencing functionality, Twitter users can address one another specifically through the @reply function—also known as a "mention"—to create a bidirectional conversation that takes form across the network. Tumasjan, Sprenger, Sandner, and Welpe (2010, p. 180) found that substantive issues can be expressed in 140 characters or fewer and that Twitter practice includes opinion expression and discussion: up to a third of all messages captured by the study were part of a conversation (p. 183). An important mechanism for organizing discussion on Twitter, of course, is the hashtag, which is user generated and bears a relationship to a discussion topic or theme. Hashtags bridge a discussion community and each member's own network (Bruns & Burgess, 2011, and Chapter 1, this volume) and catalogue a particular discussion for anyone following it.

Critical observers note a tendency on Twitter for a small cohort of prolific users to dominate discussions (Lin, Keegan, Margolin, & Lazer, 2014; Small, 2011; Tumasjan et al., 2010). They also note the tendency for Twitter posts to skew toward information broadcast rather than public dialogue (Larsson & Hallvard, 2011), and that democratic participation on Twitter tends to decline during major media events (Lin et al., 2014). With these analyses in mind, our study aims to interrogate the nature of communication activity occurring around a democratically important Canadian political event to reveal whether the medium of Twitter is well or ill suited as a discussion forum for civil society.

METHODS

This study employed Tweet Archivist to collect every tweet that included either or both of the hashtags **#ABbudget** or **#Budget2013** on and between March 5, 2013, and March 7, 2013. Tweet Archivist is an open-source online tool that allows researchers to collect all the tweets associated with a hashtag, term, or query within a specified time period and then download the tweets and associated metadata into a file for further analysis. Twitter users established **#ABbudget** to capture discussion on topics related to the 2013 Alberta budget, and the government of Alberta established and promoted **#Budget2013** on its websites (http://alberta.ca and http://budget2013.alberta.ca). In total, the data collection resulted in a total of 2,068 relevant tweets from 1,004 unique Twitter handles.

This study employed a mixed-methods approach, using quantitative content analysis to determine who was tweeting, along with critical discourse analysis to interrogate the nature of the tweets themselves. Following the work of Small (2011), Bohman (2004), and Chew and Eysenbach (2010), each tweet was manually coded for discussion and civil society properties. Both were considered as indications that participants were engaging in productive dialogue or information exchange about issues that mattered to them or to society as a whole. Additionally, the subject of each tweet was considered in order to determine if it represented information sharing or public deliberation on the issue of the 2013 budget.

INFORMATION SHARING AND SOCIAL GATEKEEPING: ALBERTA BUDGET TWITTER ACTIVITY

Who Participates?

The hashtag communities relating to the Alberta budget drew broad engagement from a wide range of actors. Of these, the high-volume participants were most noteworthy. High-volume participants in the discussion each contributed at least 3 tweets and up to 91 tweets over 3 days. On the whole, these high-volume contributors collectively added 1,654 tweets, or 62% of the total sample. Established actors, defined as Twitter accounts with affiliation to established economic, political, or social groups, contributed 35% of the total tweet bank; the most active among them were traditional mass media personalities or media outlets, which collectively contributed 21% of the total volume of tweets. Twitter accounts with no apparent affiliation to established economic, political, or social actors contributed just 28% of the total bank of 2,608 tweets. This finding shows that the majority of people involved in the discussion were not independent civic actors but were tweeting as part of their relationship to a political or social group involved in the budget. Of these, mainstream media outlets dominated the activity on this topic.

Discussion Properties

After conducting a content analysis on the sample to determine the nature of who was tweeting about the Alberta budget, we employed discourse analysis to determine the nature of the Twitter activity. To determine whether Twitter activity constituted public discussion, we drew from Small (2011) and Bruns and Burgess (2011, and Chapter 1, this volume), who suggested that a high volume of @replies and retweets that include a new comment are indicators of discussion. In our Alberta budget sample, 182 tweets were directed to at least one other Twitter user via the @reply function. Furthermore, the Twitter users in our sample used the @reply function in two distinct ways. The first was to simply name-drop or "mention" someone without the intention of engaging the other person, such as, "Watching @DougHornerMLA deliver Budget 2013 #abbudget" (2013). We excluded these types of @replies from analysis and focused instead on the use of @replies that included users mentioning another user deliberately to engage him or her in direct dialogue. This second type of @reply was deployed 108 times, accounting for 4.1% of the total sample, which is congruent with Small's (2011) study in which @replies comprised 3.1% of tweets (p. 888). An example was, "I look forward to seeing a post from [@specific user] on the #abbudget. You will be treating us to one, yes? **#ableg**" (2013). In this instance, the tweet author directly engaged a named individual and offered a specific frame for reply.

However, to users outside the specific exchange, our analysis revealed that often, conversations were opaque, at best. An example of this is seen in this tweet: "[@specific user] [@specific user] how do you figure? Experts project higher average numbers than in **#ABbudget**—would love to see your evidence..." (2013). In this tweet, the poster sought resolution to a disagreement, as demonstrated in his or her request for evidence. In tweets like this, an outside reader cannot be certain of the point with which the poster was taking issue, to what experts the poster was referring, and to what area of the budget the poster was drawing attention. Given the constraints of a 140-character message, context is at risk of abandonment.

Like @replies, modified retweets, which are retweets that include the addition of a concise editorial comment, could also be considered evidence of discussion. In our sample, only 28 tweets, or 1.1% of the data sample, were modified retweets, such as, "Some good insights MT_@[specific user] Just in time for **#ABbudget**: Beyond the Bubble @cleanenergycan http://t.co/Noo2XcW7jj **#cdnpoli #ableg #climate**" (2013). This modified retweet encourages the poster's followers to read the linked content by signaling his or her endorsement, expressed in the modified portion. It also exemplifies active engagement with content, as opposed to passive consumption.

Measured in terms of @replies and modified retweets, discussion pick-up was quite low at only 5.2% of tweets in the sample. This rate signals that participants may be broadcasting at one another rather than engaging in some type of deliberative public dialogue using the medium. Though users sought to expand their exposure by including hashtags, they lacked a commitment to monitor and respond to the ongoing discussion, a shortcoming of a platform that lacks any requirement of users to follow a discussion (Bruns & Burgess, 2011, p. 5, and Chapter 1, this volume).

Exchange of Views, Ideas, and Reasoning

Opinion expression is important in political or civil discussion because it implies reasoning, invites responses from others, and fosters exchange among different actors to challenge, accept, and examine assumptions (Bohman, 2004). Four hundred and thirteen tweets, or 15.8% of the total sample, expressed opinion, which is generally in line with Small's (2011) study that found 11.4% of the sampled tweets expressed opinion. For example, one participant wrote, "**#ABbudget** let's hope for: progressive ic tax and that the province runs a deficit w a clear plan for saving. With int rts so low, red is OK" (2013). This tweet articulates two opinions: first, the author favors tax reform shifting from a flat to progressive income tax, and second, the user outlines support for borrowing in a low-interest climate, even if the consequence is a budget deficit, qualified by an expectation for a savings plan. Such positions on tax structure and government deficits were taboo subjects in Alberta's 2013 political landscape, making this contribution important for showing that the platform and user base did not prevent its expression.

Despite the ability of the platform to support the expression of controversial opinions, we found opinion expression to be limited in our sample, which is problematic considering that public reason requires opinion exchange (Bohman, 2004; Habermas, 2006). In the Alberta budget sample, 1,270 tweets, or 49% of the sample, were neutral, coded as expressing neither a positive nor a negative opinion on a topic. Furthermore, when opinion was expressed, it was rarely debated, meaning that exchange, as defined by Castells (2008) as a conversation around views and reasoning with the purpose of reaching an opinion, rarely occurred.

Information Exchange

In a recent study of political hashtags in Canada, Small (2011) noted that Twitter users collect political information from elsewhere on the Internet and share it with their followers using hashtags. In her study, 47% of examined tweets shared links to online information (p. 884). Twitter users in our study were also relatively active with sharing online information. We observed that 542 posts, or 20.8% of the Alberta budget sample, shared information directly through attached hyperlinks. The different results between the two studies may be due to data skewing resulting from the presence of two political blog aggregators in Small's (2011) data that automated the dissemination of blog post titles and links (p. 882), facilitating, though importantly, not participating in, public dialogue. Such aggregator activity was not apparent in the Alberta budget study's data, likely because it examined a time-specific hashtag community, whereas Small's examined an ongoing hashtag community.

There is thus a culture of information sharing and a commitment to equal access evidenced in the Alberta budget sample; however, the direction of information flows is important in evaluating whether information sharing itself can be considered participatory public dialogue, and here the influence and effects of the high-volume Twitter participants matter a great deal. When individuals did use the #ABbudget or #Budget2013 hashtags to share links to information located elsewhere on the web, there were clear patterns in the types of sites that were hyperlinked to. In our sample, hyperlinks most often pointed to traditional media resources (321 instances, or 59.2%), civic resources (130 instances, or 24.0%), alternative media (57 instances, or 10.5%), partisan resources (29 instances, or 5.4%), and labour resources (3 instances, or 0.6%). Thus, the particular hyperlinks shared in our sample most often directed other Twitter users to broadcast media outlets elsewhere on the web, demonstrating a particular use of Twitter as a promotional tool for traditional media actors.

Twitter: A Participatory Culture That Amplifies Power

As demonstrated above, news media actors from established news organizations drove the flow of content in at least two important ways. First, the amplification of established voices attained through retweet activity was significant. Twitter users redistributed 1,163 tweets, or 44.6% of the data set, as retweets, including retweets redistributed with an editorial comment; 706 of those messages, or 27.1% of the data set, originated from accounts controlled by established actors, including traditional media, government, politicians, labour unions, and political parties. At 555, or 21.3%, the majority of retweets originated with traditional media actors. This is important because such actors already control powerful platforms from which to communicate and influence the general public and public opinion. When Twitter is used in this way, it serves primarily as an echo chamber for those ideas already in the popular press, further problematizing the critiques leveled by such scholars as Tumasjan et al. (2010), Small (2011), and Larsson and Hallvard (2011). Furthermore, our content analysis reveals that this culture of redistributing tweets served to repeat and amplify the messages of those in power rather than critique them, e.g., "RT @ctvedmonton: **#ABbudget**: Alberta's publicly funded post secondary institutions will get $2 billion in base operating grants. **#yeg #ableg**" (2013). In this example, the absence of context or a benchmark belies that base operating grants to institutions were actually reduced by 7.3%, or $145 million, in the 2013 Alberta budget from the year previous. In the Twitter discussion of the Alberta budget, then, we see not a venue for public dialogue, but a new distribution platform for traditional media influencers and their uncontested points of view.

Complicating the scenario, some tweets related to **#ABbudget** seemed to be designed to direct people away from Twitter in favor of less dialogic media platforms. In our sample, tweets from news radio programming or television news actors were constructed in such a way as to drive Twitter discussion participants to their respective schedule- and platform-dependent broadcasts, e.g., "Also on CTV News at Noon: @BillFortierCTV has a preview of the **#ABbudget**, which will be tabled tomorrow. **#yeg #CTVYEG**" (2013). This illustrates Jenkins's (2006) notion of media convergence. Consumers are prompted to seek out information dispersed over several platforms and from sources promoting themselves as trusted.

The direction of users to traditional media sites is a strong example of how the public and private spheres are interwoven on Twitter. Consider this tweet posted by a traditional media actor, GlobalEdmonton: "Join the Alberta Budget conversation on our live blog by using **#GNlive** http://t.co/tNUpTzrMCi **#abbudget #yeg #yyc**" (2013). Here, the outlet created a live blog and associated hashtag that served to fragment the discussion community. If users followed the link to the live blog, the loss for the Twitter community was at least twofold: first, the user would potentially no longer be following and reacting to the discussion there, and second, those posts contributed directly to the live blog would not be simultaneously posted back on Twitter where they could contribute to the general discussion. In this case, Twitter acts as an entry point to a walled garden, controlled by powerful traditional media outlets that seek to gain eyeballs and content production from a commodified audience (Smythe, 1981; Van Dijck, 2013). A conversation on Twitter does not create revenues for news media outlets as it does when hosted on their own platforms where their advertising placements reside (Warner, 2013). In our study, we observed the private interests of news media driving their agents' participation on Twitter and, in turn, driving the discussion on Twitter in general.

Literature on the general nature of online participation elicits concepts of horizontal networks of citizen actors engaged in discourse, bypassing media and government control (Castells, 2008, p. 90; Drache, 2008; Shirky, 2008), and the many contributing to agenda setting and issues rising to public salience (Benkler, 2006, p. 204). This study's findings diverge from these notions. The data show traditional media actors driving discussion and enjoying significant amplification from the general user base, while general users add very little in the way of critique or dialogue. With the means of dissemination so accessible to the masses, the associated outcomes seem nevertheless to elicit a mass participatory culture intent on reiterating the message of those in power—far from the functional notions of active citizens setting the terms of discussion and driving it away from, or keeping it in some kind of balance with, traditional gatekeepers. Unfortunately, with the recent Twitter IPO, platform redesign, and associated need to make money for a new set of shareholders, we fear that this situation will only continue to worsen, with Twitter itself becoming an advertiser-supported walled garden to rival even the most established traditional media outlets.

WHEN SOME VOICES ARE AMPLIFIED: PROBLEMATIZING MEDIA DOMINANCE ON TWITTER

This study examined the potential for Twitter hashtag communities to serve as discussion spaces for civil society. In total, 2,608 tweets were read and manually coded as part of the study. Overall, for a topic of great public import, discussion pick-up on Twitter seemed quite low. Indicators of dialogue or discussion, as measured by @replies and retweets with additional and new commentary, were relatively scant. It was not clear, then, if tweets were posted with an expectation of response. Importantly, this study noted that @replies tend to abandon context and diminish their publicness by consequence. Opinion expression was also not particularly strong, with just 413 tweets, or 15.8% of the sample, expressing opinion as outlined by Small (2011). In contrast, however, the **#ABbudget** hashtag community seemed particularly keen to exchange information, including rich information (e.g., multimedia). Indeed, 542 tweets included hyperlinks, the majority of which pointed to traditional media content. This information sharing was actually indicative of a separate trend witnessed in the **#ABbudget** hashtag community: the use of Twitter by traditional media actors as a venue for attracting additional audience members back to a primary website or other medium.

Traditional media have embraced Twitter as a way to quickly share news, drive traffic to their sites, and, as Warner (2013) observed, build relationships and trust with consumers to do the thinking for them (p. 268). To this end, traditional media actors dominated the Twitter discussion, an accepted part of the Twitter culture given amplification of those in power and generous sharing of traditional media content. Traditional media actors dominated the **#ABbudget** Twitter discussion and did so in a broadcast rather than dialogic way—not supporting public discussion but potentially undermining it. Rather than serving as a source of information to inspire online public debate on this platform, these dominant media actors instead, in the case of the **#ABbudget** hashtag community, served to fragment the discussion community by driving participants to news outlet sites and perspectives deemed important by the news media.

We are at a critical juncture: on one hand, there is this space that could enable a networked public to exchange views and design and articulate their own aspirations (Shirky, 2008), and on the other, there are old actors who are unwilling to relinquish their role as powerful drivers of public opinion contesting this same space (Drache, 2008). The challenge, as articulated by Jenkins (2006), is "whether the public is ready to push for greater participation or willing to settle for the same old relations to mass media" (p. 243). While there is evidence that diverse actors contributed to the formation of the Twitter **#ABbudget** hashtag community, the associated discussion failed to meet the functional characteristics of the public sphere. The discussion lacked an exchange of diverse ideas, perspectives, and opinion—the basis for a community to coalesce around compelling ideas and, in turn, articulate shared aspirations to those controlling the levers of power. Furthermore, the **#ABbudget** discussion showed a dominance of mainstream media influencers and the formation of an "online echo chamber" (Larsson & Hallvard, 2011), which served to direct people to less dialogic news media elsewhere. Therefore, our findings are congruent with Lin et al.'s (2014) study that showed that the discussion of major issues on Twitter tends to amplify already dominant voices and does little to create democratic exchange.

Twitter's most apparent value to civil society is information exchange in terms of both tweet content and hyperlinked content. However, this study suggests that the affordances of Twitter do not, in and of themselves, encourage an online public sphere (Benkler 2006; Drache, 2008). The crucial component of public dialogue—the articulation of opinion—may actually be quite challenging in 140 characters or less, particularly for a mundane but very important matter like a provincial budget. Furthermore, the presence of a small group of users dominating the online discussion (Tumasjan et al., 2010) was shown to indeed be a problem for public discourse on this issue, and one that is not mitigated by the presence of hashtags to organize the Twitter information stream. This suggests that Twitter should not be held up as a potential platform for the online public sphere. We should recognize that, on Twitter, some users and perhaps some topics are more likely to inspire dialogue than others, and a hashtag community in and of itself does not represent dialogue on a given issue.

REFERENCES

Benkler, Y. (2006). *The wealth of networks: How social production transforms markets and freedom*. New Haven, CT: Yale University Press.

Bohman, J. (2004). Expanding dialogue: The Internet, the public sphere, and prospects for transnational democracy. *Sociological Review, 52*(1), 131–155. doi: 10.1111/j.1467-954X.2004.00477.x

Bruns, A., & Burgess, J. (2011). *The use of Twitter hashtags in the formation of ad hoc publics*. Paper presented at the 6th European Consortium for Political Research General Conference, August 25–27, University of Iceland, Reykjavik.

Castells, M. (2008). The new public sphere: Global civil society, communication networks, and global governance. *Annals of the American Academy of Political and Social Science, 616*(1), 78–93. doi: 10.1177/0002716207311877

Chew, C., & Eysenbach, G. (2010). Pandemics in the age of Twitter: Content analysis of tweets during the 2009 H1N1 outbreak. *PLoS ONE 5*(11), 1–13. doi:10.1371/journal.pone.0014118

Drache, D. (2008). *Defiant publics: The unprecedented reach of the global citizen*. London: Polity.

Gerhards, J., & Schäfer, M. S. (2010). Is the Internet a better public sphere? Comparing old and new media in the USA and Germany. *New Media & Society, 12*(1), 143–160. doi: 10.1177/1461444809341444

Habermas, J. (2006). Political communication in media society: Does democracy still enjoy an epistemic dimension? The impact of normative theory on empirical research. *Communications Theory, 16*(4), 411–426. doi:10.1111/j.1468-2885.2006.00280.x

Jenkins, H. (2006). *Convergence culture: Where old and new media collide*. New York: New York University Press.

Kim, Y., Hsu, S. H., & Gil de Zúñiga, H. (2013). Influence of social media use on discussion network heterogeneity and civic engagement: The moderating role of personality traits. *Journal of Communication, 63*(3), 498–516. doi: 10.1111/jcom.12034

Larsson, A. O., & Hallvard, M. (2011). Studying political microblogging: Twitter users in the 2010 Swedish election campaign. *New Media & Society, 14*(5), 729–747. doi: 10.1177/1461444811422894

Lin, Y., Keegan, B., Margolin, D., & Lazer, D. (2014). Rising tides or rising stars? Dynamics of shared attention on Twitter during media events. *PLoS ONE 9*(5). Retrieved from http://journals.plos.org/plosone/article?id=10.1371/journal.pone.0094093

Pariser, E. (2011). *The filter bubble: What the Internet is hiding from you.* New York: Penguin.

Shirky, C. (2008). *Here comes everybody: The power of organizing without organizations.* New York: Penguin.

Small, T. (2011). What the hashtag? A content analysis of Canadian politics on Twitter. *Information, Communication & Society, 14*(6), 872–95. doi: 10.1080/1369118X.2011.554572

Smythe, D. (1981). *Dependency road: Communications, capitalism, consciousness, and Canada.* Norwood, NJ: Ablex.

Tumasjan, A., Sprenger, T. O., Sandner, P. G., & Welpe, I. M. (2010). Predicting elections with Twitter: What 140 characters reveal about political sentiment. *Proceedings of the Fourth International AAAI Conference on Weblogs and Social Media (ICWSM),* 2010.

Twitter. (2014). *Fact sheet.* San Francisco

Van Dijck, J. (2013). *The culture of connectivity.* New York: Oxford University Press.

Warner, J. (2013). The new refeudalization of the public sphere. In M. P. McAllister & E. West (Eds.), *The Routledge companion to advertising and promotional culture* (pp. 285–297). New York: Routledge.

CHAPTER TWENTY

Hashtagging #HigherEd

SAVA SAHELI SINGH

Today's academic landscape is changing significantly and rapidly, and scholars entering academia have to overcome bigger and different hurdles—jobs are more competitive, expectations higher, the need to stand out more urgent. Scholars have begun to use the Internet as a place to build and create academic identities to supplement their CVs and as a way to showcase technology skills while establishing an online presence and identity—valuable traits in an evolving academic world. This can take many forms, including, for example: personal blogs, online journals, mainstream online outlets, a page on academia.edu, or a social media presence.

In this chapter, I will cover some of the ways in which hashtags on Twitter contribute to building and maintaining academic identity, such as promoting one's own work by using hashtags, partaking in hashtag communities, and reaping the benefits of membership in a networked community of academics. I will also discuss how Twitter hashtags are used in the classroom and how academic research has embraced social media platforms such as Twitter because of the way data can be mined. It is important to keep in mind, however, that there are possible negative effects of using social media platforms generally and hashtags specifically, such as creating echo chambers and cliques, and causing pigeonholing.

#CONFERENCES

Twitter plays an interesting role for early-career academics or Ph.D. students making their first forays into the job market. Aside from community and commiseration, Twitter also provides a place for academics to find connections and collaborators, and hashtags can be key to making this possible. Hashtags provide an efficient way to parse and filter information and people on Twitter, making it easier to follow conversations, trends, and people. A good example of this is hashtag use at conferences.

Conferences are great for building academic community and networking, and in many cases, academics are more likely to use Twitter during conferences (Reinhardt, Ebner, Beham, & Costa, 2009; Ross, Terras, Warwick, & Welsh, 2011). The connections made at conferences can change the course of your academic life. But not everyone can make it to conferences—a lot depends on department funds, personal funds, and time, among other things. Those who can't attend have to be content with following the conference hashtag online—now a fixture of most, if not all, conferences. Those who do attend play the role of reporters: they send out vignettes of conference proceedings, hungrily waiting for the perfect soundbite that will fit 140 characters. Ross et al. (2011) refer to some conference goers as "social reporters"—people who tweet notes from a conference for those who are not there, like collective note-taking. The hashtag also serves to identify other attendees, creating a spontaneous, short-term community that can use the hashtag to contextualize a conference backchannel, a way to have a conversation about conference proceedings that isn't in the forefront.

Those who attend also use conference hashtags to identify others who are present—either interesting colleagues to network with or old friends to get a drink with; and to identify themselves as present, to both those in attendance and those following from afar. Doing so immediately confirms the tweeter's membership to the community that is represented by the conference and hashtag, thus increasing visibility within the community. Conference hashtags are also a great way to identify and follow other members of this community.

Before the mass uptake of Twitter, conference hashtags served mainly as a backchannel at conferences. A few attendees could use the space without much scrutiny or visibility in order to comment on conference proceedings to engage with presenters, debate some of the finer points with each other, or even snark at presenters and presentations. Now, with Twitter's increasing popularity, it is less a backchannel and more a parallel signaling stream and announcement center, with many conferences setting up big screens to display a live stream of conference tweets. Along with this increased visibility, there is an increase in the need to be visible, to be identified as part of the community.

#COMMUNITIES

Because hashtags help to contextualize tweets and Twitter conversations, one can tag tweets aimed at particular groups of people. For example, if I tweet a link to an article about educational technology and tag my tweet with **#edtech**, a few things can happen:

- people who are interested in educational technology and have **#edtech** loaded in a column in their Twitter application of choice or saved as a search will see my tweet;
- people who follow me will see the hashtag and understand that the content has something to do with educational technology;
- those of my followers interested in educational technology will pay it some attention, while those not interested in educational technology can ignore the tweet and its contents;
- if my tweet is retweeted, people who do not follow me will see the tweet and react to it in one of the above ways, and they will also associate me with that kind of content, which could lead to them following me.

Because hashtags help to contextualize tweets and can be used to index and search Twitter, this logically extends to community cohesion. Academic hashtag communities on Twitter can take different forms. They can be loose interest groups that share resources and information or more organized communities with specific meeting times and community involvement beyond Twitter. Hashtags can identify a user to a particular community to show solidarity or share resources (e.g., **#academicboycott** and **#phdlife**), or they can aim to introduce networks to each other or highlight one's network to others, as does **#ScholarSunday** (Pacheco-Vega, 2012), which some scholars use as a kind of academic **#FollowFriday**.

Academic hashtag communities often form around support for academic processes. **#phdchat** is a popular hashtag started by Nasima Riazat (2011), a scholar based in the U.K. It is mostly frequented by potential, current, or recent Ph.D. students and academics with an interest in mentoring and connecting with Ph.D. students. **#phdchat** is a great example of a hashtag community because it follows a simple and effective process and speaks to a very important part of academic life. An online poll is circulated every week with a list of topics, members of the community vote on the topic they'd like to discuss that week, and when the day for **#phdchat** rolls around, the moderator—often Riazat herself—starts the "chat" off for the hour by announcing the topic (anything ranging from tips on how to write to surviving institutional hurdles) and inviting others to join in using the hashtag. Academics and scholars from all over the world then contribute to the conversation by responding to the prompt, with either questions they have or answers to the questions posed by other members of the community. These chats are often storified or saved, either on a wiki or on a community member's website, so that they can be accessed later for reference.

In the time I've participated in and observed this community, a definite structure was apparent. Lave and Wenger's (1991) concept of a community of practice (CoP) maps well onto how this community works. Senior scholars, faculty members, and people who have been part of #phdchat for longer tend to guide newcomers, moving them from legitimate peripheral participation (Lave & Wenger, 1991) to those with a more central place—both within the community and in their own knowledge. In an ephemeral and fast-moving site like Twitter, what keeps the community together is its ongoing need for community and willingness to share in that community. As Ford, Veletsianos, and Resta (2014, p. 18) find in their analysis, "#PhDchat represents the desires and needs of its members, and its ability to disseminate information is key to its mechanisms for sustaining itself." Outside of running the weekly chats, the hashtag also serves as a marker for resources and content pertaining to the Ph.D. community; members continue to share resources and articles throughout the week, appending each tweet with #phdchat to let other members know it might be of interest.

#acwrimo, for "academic writing month," is an offshoot of #nanowrimo, for National Novel Writing Month, which usually falls in November. The non-month version of #acwrimo is just #acwri, for "academic writing," which academic communities use for the specific task of scholarly writing. Scholars and academics on Twitter use this latter hashtag to denote that they are writing. It is a way to motivate groups of people on Twitter to do writing sprints or pomodori,[1] or a way to self-motivate and keep track of their own writing. Using the hashtag and announcing a word count often elicits praise, and if someone is having a hard time, community members offer encouragement.

#CLASSROOMS

Studies have looked at how Twitter is used in second-language learning (Borau, Ullrich, Feng, & Shen, 2009), marketing classes (Lowe & Laffey, 2011; Rinaldo, Tapp, & Laverie, 2011), and as a backchannel for recursive writing and collaborative meaning making through tweeting (McNely, 2009). When used as a backchannel in the classroom, it has been shown to be useful in large classes where students don't often get to engage with each other or the instructor. In this setting, Twitter gives them a voice that they otherwise might not have had, a way to express themselves if they are not comfortable with speaking up in class.

Some instructors have used Twitter effectively outside of the classroom to engage with students as they complete assignments for class. Instructors sometimes post prompts that students have to respond to by tweeting. Dunlap and Lowenthal (2009) use Twitter for just-in-time social interactions and connections where students can access professors (and vice versa) outside academic office hours or class time. Johnson (2011) links Twitter behavior outside the class setting with instructor credibility; and Junco, Heiberger, and Loken (2010) link Twitter use in a college classroom with higher levels of engagement and better grades. As with conferences and communities, using Twitter in this way allows instructors to track activity in a class community and to be accessible to students outside of a classroom setting.

For many online courses and massively open online courses (MOOCs), Twitter forms an important communication node for the community of online students. Not all MOOCs are tied to proprietary software platforms, and, in fact, for earlier MOOCs (often referred to as cMOOCs in some circles, with the "c" standing for "connectivist") (Downes, 2013), Twitter was one of the main sites for conversation and connection for online students. Even with the onslaught of the proprietary MOOCs and their built-in communication, Twitter often still plays some part in the learning community's communication. In their analysis of two German MOOCs, van Treeck and Ebner (2013) found that while Twitter use seems variable and often concentrated around a smaller percentage of users, it is still a relevant platform for communication.

#ds106 (http://ds106.us/) is a good example of an online course that went on to create ongoing community outside and after the online course and continues to connect participants from all over the world. It started as a digital storytelling class taught by Jim Groom at the University of Mary Washington and became a global community of learners and teachers. #ds106 is set up as a class at the university, but what Groom did was to open it up to anyone who wanted to join. Those students who had enrolled in the class for credit had to complete all the assignments, but open participants could dip in and out of the class and assignments as they pleased. This created a vast network of participants who then created a huge corpus of content—blog posts, videos, photos, gifs, tweets, sound clips—all using free and online applications and platforms. Opening up the class allowed students to learn not only the content of the course but also ways of being on the Internet and connecting with other people through their work.

In all of the cases above, educators use Twitter as a platform for communication and connection, and also as a way to help students learn, as in #ds106, about "being on the Internet." It is a way to get students to think about digital citizenship and online behavior in their own contexts, and a place to practice it in a way that is scaffolded by their community and instructor.

#IDENTITY

Twitter provides the space for people to create identities and enact them for an imagined audience or even a networked audience (Marwick & boyd, 2010). Creating an online presence or brand has become almost as important as building a robust CV. Often, the first thing people do after they've met someone online, or even during a job search, is search for them on the Internet. Along with personal or professional websites, social media results show up as part of a person's online identity.

Users hashtag their tweets in order to "increase the visibility of their own messages, even if they do not themselves track the hashtagged tweets" (Bruns & Moe, 2014, p. 18). Hashtags have become important identity units, and creating the right combination of units has become important for careers and connections. Being identified with a particular community means one gets to avail of what that community has to offer: jobs, writing opportunities, collaborations, and more. A prominent node in a community can become an important ally—not just as a contact but also by association.

For example, conference hashtags can establish that a person is part of the community or discipline connected to particular conferences. If that same person uses **#acwri** and reports on their writing progress often, it gives the perception of the person being a "serious" scholar. If they partake in open online courses using the course hashtags, then they are seen as being active in their own and others' learning. If they are teaching a class and using Twitter, they might be seen as innovative and forward-thinking. One could support online activism by using the appropriate hashtag, and that could signal an interest in the welfare of one's community. All this adds up to create the type of profile that potential collaborators and employers would regard as favorable.

Academics who use Twitter can increase their influence and build a personal brand that they can then leverage into building a more desirable public profile. Aligning with specific groups of people and fields of study ensures that they are associated with those communities, and those communities are more often than not formed around hashtags. This makes it easier for them to move through that community or group of people and increase their visibility, because people watching will know or believe that they are part of that community. This allows people to be "pre-vetted" in a way and to build social capital and legitimacy through social association. This network of "strong weak ties" (Granovetter, 1973) makes it possible for people in a network to share resources and opportunities, possibly providing a head start in the job market.

#NOTALLHASHTAGS

Not all uses of hashtags in academic settings have positive implications, however. While creating a visible identity ensures attention and presence, it can also lock someone into that identity. Aligning with certain communities creates silos of people that are not as open to other communities or individuals. These communities run the risk of creating cliques that might be perceived as hostile or difficult to penetrate (Singh, 2013).

There are also some hashtags that are born of controversy. For example, in 2012, there was a robust and sometimes contentious discussion around the ethics and etiquette of using Twitter during conferences. People adopted **#Twittergate** as the identifying hashtag and fell on one side or the other of the debate (Cottom, 2012). Such occurrences can cause rifts in otherwise peaceful communities. Moreover, some activist hashtags might be politically divisive, and, at worst, can result in the loss of jobs (Schmidt, 2014). As for the impact on students, those who deploy Twitter in the classroom or for online courses often encourage but don't require its use. If people choose not to use it, they risk missing out on some aspects of the learning experience and limiting their access to the broader community. Hashtags can also be irksome—people might not appreciate other people's enthusiastic tweets and hashtags for things like national sporting events or popular TV shows. Finally, hashtags like **#GamerGate** can bring a torrent of abuse and threats against people just for disagreeing with a particular position (Wingfield, 2014). As much as hashtags can bring people together, they can also tear communities apart.

#RESEARCH

In Twitter, researchers have found a proxy for studying human beings, and most of these phenomena unfold around hashtags, which makes it very easy to mine and gather data, analyze it using software, and visualize it for a better understanding of trends. The amount of scholarship surrounding Twitter increases every day; studies cover cultural events in television and film (Buschow, Schneider, & Ueberheide, 2014; Harrington, 2013), politics (Maireder & Ausserhofer, 2013; Meraz & Papacharissi, 2013), cultural phenomena and memes (Leavitt, 2013), activism (Conley, 2014; Loza, 2014), and race (Florini, 2013; Sharma, 2013), to name a few, and hashtags play a central role in each of these studies.

A good indicator of how much of a platform for research Twitter has become is the number of articles laying out methods for data collection and analysis (e.g., Beurskens, 2013; Bruns & Stieglitz, 2013; Einspänner, Dang-Ahn, & Thimm, 2013; Gaffney & Puschmann, 2013; Marwick, 2013; Thelwall, 2013). In fact, the book *Twitter and Society* (Weller, Bruns, Burgess, Mahrt, & Puschmann, 2013) is a recent comprehensive collection of articles (including the ones listed in the previous sentence) that cover a breadth of topics related to Twitter, within which hashtags form a central element of study.

Weller and colleagues (2013) call for the continued study of Twitter because "[Twitter] provides a window on contemporary society as such, at national and global levels" (p. 426), and as Bruns and Moe point out in the same volume, hashtags remain the most straightforward way to interrogate Twitter because they are the most easily tracked phenomenon on Twitter, and because "[M]ethodologically, it is considerably more difficult to move beyond the relatively well-behaved confines of macro-layer hashtag studies" (2013, p. 25).

#CONCLUSION

The sections covered in this chapter provide insight into some of the facets of hashtags, the roles they play in the lives of academics and scholars on Twitter, and how this little annotation has had a significant impact on culture, communities, and communication.

As with most social media platforms, Twitter and hashtags work for those who invest time in them. People can be motivated to tweet by identifying with communities because they see it as integral to building an online presence or because a friend or colleague suggested they try it. Either way, more and more people are aware of what Twitter is and the power it can have, good and bad. But can it scale? As Twitter has grown in popularity, the content is shifting toward more marketing and advertising, and relevant content is tussling with irrelevant content for attention. As hashtag communities grow, it is yet to be seen if Twitter as a platform will be able to support the volume of users, or if users will start seeking out alternative platforms on which to recreate their communities.

While it is still early to guess at its eventual extent, we cannot ignore Twitter's impact on academia. It is obvious that Twitter and hashtagging are changing the nature of current academic practices and expanding how academics build communities, identities, and scholarship that challenge what it means to be an academic in a digital age.

NOTE

1. The pomodoro technique is a method to manage time by breaking tasks up into 25-minute increments, or pomodori. It was developed by Francesco Cirillo in the 1980s. http://pomodoro technique.com/

REFERENCES

Beurskens, M. (2013). Legal questions of Twitter research. In K. Weller, A. Bruns, J. Burgess, M. Mahrt, & C. Pushmann (Eds.), *Twitter and society* (pp. 123–133). New York: Peter Lang.

Borau, K., Ullrich, C., Feng, J., & Shen, R. (2009). Microblogging for language learning: Using Twitter to train communicative and cultural competence. In *Advances in Web Based Learning–ICWL 2009* (pp. 78–87). Berlin, Heidelberg: Springer.

Bruns, A., & Moe, H. (2013). Structural layers of communication on Twitter. In K. Weller, A. Bruns, J. Burgess, M. Mahrt, & C. Pushmann (Eds.), *Twitter and society* (pp. 15–28). New York: Peter Lang.

Bruns, A., & Stieglitz, S. (2013). Metrics for understanding communication on Twitter. In K. Weller, A. Bruns, J. Burgess, M. Mahrt, & C. Pushmann (Eds.), *Twitter and society* (pp. 69–82). New York: Peter Lang.

Buschow, C., Schneider, B., & Ueberheide, S. (2014). Tweeting television: Exploring communication activities on Twitter while watching TV. *Communications—The European Journal of Communication Research, 39*(2), 129–149.

Conley, T. L. (2014). From #RenishaMcBride to #RememberRenisha: Locating our stories and finding justice. *Feminist Media Studies, 14*(6), 1111–1113.

Cottom, T. M. (2012, September 30). *An idea is a dangerous thing to quarantine #twittergate* [Blog post]. Retrieved from http://tressiemc.com/2012/09/30/an-idea-is-a-dangerous-thing-to-quarantine-twittergate/

Downes, S. (2013). *Massively open online courses are "here to stay"* [Blog post]. Retrieved from http://www.downes.ca/post/58676

Dunlap, J. C., & Lowenthal, P. R. (2009). Tweeting the night away: Using Twitter to enhance social presence. *Journal of Information Systems Education, 20*(2).

Einspänner, J., Dang-Ahn, M., & Thimm, C. (2013). Computer-assisted content analysis of Twitter data. In K. Weller, A. Bruns, J. Burgess, M. Mahrt, & C. Pushmann (Eds.), *Twitter and society* (pp. 98–108). New York: Peter Lang.

Florini, S. (2013, March 7). Tweets, Tweeps, and Signifyin': Communication and Cultural Performance on "Black Twitter." *Television & New Media.*

Ford, K., Veletsianos, G., & Resta, P. (2014). The structure and characteristics of #PhDChat, an emergent online social network. *Journal of Interactive Media in Education, 18*(1). Retrieved from http://jime.open.ac.uk/article/view/2014-08/533

Gaffney, D., & Puschmann, C. (2013). Data collection on Twitter. In K. Weller, A. Bruns, J. Burgess, M. Mahrt, & C. Pushmann (Eds.), *Twitter and society* (pp. 55–67). New York: Peter Lang.

Granovetter, M. S. (1973). The strength of weak ties. *American Journal of Sociology, 78*, 1360–1380.

Harrington, S. (2013). Tweeting about the telly: Live TV, audiences, and social media. In K. Weller, A. Bruns, J. Burgess, M. Mahrt, & C. Pushmann (Eds.), *Twitter and society* (pp. 237–247). New York: Peter Lang.

Johnson, K. A. (2011). The effect of Twitter posts on students' perceptions of instructor credibility. *Learning, Media and Technology, 36*(1), 21–38. doi: 10.1080/17439884.2010.534798

Junco, R., Heiberger, G., & Loken, E. (2011). The effect of Twitter on college student engagement and grades. *Journal of Computer Assisted Learning, 27*(2), 119–132. doi: 10.1111/j.1365-2729.2010.00387.x

Lave, J., & Wenger, E. (1991). *Situated learning: Legitimate peripheral participation*. London: Cambridge University Press.

Leavitt, A. (2013). From #FollowFriday to YOLO: Exploring the cultural salience of Twitter memes. In K. Weller, A. Bruns, J. Burgess, M. Mahrt, & C. Pushmann (Eds.), *Twitter and society* (pp. 137–154). New York: Peter Lang.

Lowe, B., & Laffey, D. (2011). Is Twitter for the birds?: Using Twitter to enhance student learning in a marketing course. *Journal of Marketing Education, 33*(2), 183–192. doi: 10.1177/0273475311410851

Loza, S. (2014). Hashtag feminism, #SolidarityIsForWhiteWomen, and the other #FemFuture. *Ada: A Journal of Gender, New Media, and Technology*, (5). Retrieved from http://adanewmedia. org/2014/07/issue5-loza/

Maireder, A., & Ausserhofer, J. (2013). Political discourses on twitter: Networking topics, objects, and people. In K. Weller, A. Bruns, J. Burgess, M. Mahrt, & C. Pushmann (Eds.), *Twitter and society* (pp. 305–318). New York: Peter Lang.

Marwick, A. E. (2013). Ethnographic and qualitative research on Twitter. In K. Weller, A. Bruns, J. Burgess, M. Mahrt, & C. Pushmann (Eds.), *Twitter and society* (pp. 109–121). New York: Peter Lang.

Marwick, A., & boyd, d. (2011). I tweet honestly, I tweet passionately: Twitter users, context collapse, and the imagined audience. *New Media and Society, 13*, 96–113.

McNely, B. (2009). *Backchannel persistence and collaborative meaning-making*. Paper presented at the 27th ACM International Conference on Design of Communication, October 5–7, Bloomington, IN.

Meraz, S., & Papacharissi, Z. (2013, January 27). Networked gatekeeping and networked framing on #Egypt. *The International Journal of Press/Politics*. doi: 10.1177/1940161212474472

Pachego-Vega, R. (2012, September 2). *Using the #ScholarSunday hashtag as a #FollowFriday for academics* [Blog post]. Retrieved from http://www.raulpacheco.org/2012/09/scholarsunday/

Reinhardt, W., Ebner, M., Beham, G., & Costa, C. (2009). How people are using Twitter during conferences. In V. Hornung-Prähauser & M. Luckmann (Eds.), *Creativity and innovation competencies on the Web. Proceeding of 5 EduMedia Conference* (pp. 145–156). Salzburg, Austria.

Riazat, N. (2011, August 10). *A doctoral networking forum: the origins of #phdchat* [Blog post]. Retrieved from http://phdchat-nasimariazat.blogspot.com/2011/08/nasima-riazat-doctoral-networking-fo rum.html

Rinaldo, S. B., Tapp, S., & Laverie, D. A. (2011). Learning by tweeting: Using Twitter as a pedagogical tool. *Journal of Marketing Education, 33*(2), 193–203.

Ross, C., Terras, M., Warwick, C., & Welsh, A. (2011). Enabled backchannel: Conference Twitter use by digital humanists. *Journal of Documentation, 67*(2), 214–237.

Schmidt, P. (2014, August 7). Denial of job to harsh critic of Israel divides advocates of academic freedom. *The Chronicle of Higher Education*. Retrieved from http://m.chronicle.com/article/Denial-of-Job-to-Harsh-Critic/148211/

Sharma, S. (2013). Black Twitter?: Racial hashtags, networks and contagion. *New Formations: A Journal of Culture/Theory/Politics, 78*(1), 46–64.

Singh, S. S. (2013). Tweeting to the choir: Online performance and academic identity. *Selected Papers of Internet Research, 13*. Retrieved from http://spir.aoir.org/index.php/spir/article/view/835

Thelwall, M. (2013). Sentiment analysis and time series with Twitter. In K. Weller, A. Bruns, J. Burgess, M. Mahrt, & C. Pushmann (Eds.), *Twitter and society* (pp. 83–95). New York: Peter Lang.

van Treeck, T., & Ebner, M. (2013). How useful is Twitter for learning in massive communities? An analysis of two MOOCs. In K. Weller, A. Bruns, J. Burgess, M. Mahrt, & C. Pushmann (Eds.), *Twitter and society* (pp. 411–423). New York: Peter Lang.

Weller, K., Bruns, A., Burgess, J. E., Mahrt, M., & Puschmann, C. (Eds.). (2013). *Twitter and society*. New York: Peter Lang.

Wingfield, N. (2014, October 15). Feminist critics of video games facing threats in "Gamergate" campaign. *The New York Times*. Retrieved from http://www.nytimes.com/2014/10/16/technology/gamergate-women-video-game-threats-anita-sarkeesian.html?_r=0

#Contributors

Geane Alzamora is an adjunct professor in the Social Communication Department at the Federal University of Minas Gerais, Brazil. She investigates new approaches in theory of communication based on semiotics and actor-network theory, specifically related to topics such as transmedia and social mobilization in the interface between the streets and the screens. Her research related to the topic of this paper is supported by Capes, CNPq, and Fapemig. E-mail: geanealzamora@ufmg.br

Carlos d'Andréa is a professor in the Social Communication Postgraduate Program at the Federal University of Minas Gerais, Brazil. Based in Actor-Network Theory and Complex Adaptive Systems theory, he currently researches the emergence of mediatized controversies in intermedia connections between television and online social networks. He is also interested in new methods (such as data scraping and visualizations) for studying digital media. His research is supported by CNPq and Fapemig. E-mail: carlosfbd@gmail.com

Anna Antonakis-Nashif is a Ph.D. student at the Freie University, Berlin. She holds a master's degree in political science and has acquired regional expertise on Tunisia via fieldwork visits since 2009. Her research revolves around mobilization dynamics at the interface of on- and offline and includes intersectional theories and theories of (counter)publics. In her Ph.D., she analyzes the creation and political impact of feminist counterpublic spheres in Tunisia after the uprisings in 2011.

Brett Bergie is a graduate of the M.A. professional communication program at Royal Roads University and works in communications and strategic planning at Bow Valley College in Calgary, Canada. Her research interests are concerned with the formation and political behaviors of online publics as well as diverse gender expression and social inclusion. She gathers strength, clarity, and amusement from Beatrice and Samuel. E-mail: bbergie@bowvalleycollege.ca; Twitter: @brettbergie

Stacy Blasiola is a doctoral candidate in the Department of Communication at the University of Illinois at Chicago and the recipient of a National Science Foundation IGERT Fellowship in Electronic Security and Privacy. Her research examines political activism, platforms, algorithms, and privacy.

Axel Bruns is an Australian Research Council Future Fellow and professor in the Creative Industries Faculty at Queensland University of Technology in Brisbane, Australia. He is the author of *Blogs, Wikipedia, Second Life and Beyond: From Production to Produsage* (2008) and *Gatewatching: Collaborative Online News Production* (2005), and a co-editor of *Twitter and Society* (2014), *A Companion to New Media Dynamics* (2012), and *Uses of Blogs* (2006). Web: http://snurb.info/; Twitter: @snurb_dot_info

Jean Burgess is director of the Digital Media Research Centre at Queensland University of Technology. She researches the everyday uses and politics of social and mobile media platforms, as well as new digital methods for studying them. Her books include *YouTube* (Polity Press, 2009), *Studying Mobile Media* (Routledge, 2012), *A Companion to New Media Dynamics* (Wiley-Blackwell, 2013), and *Twitter and Society* (Peter Lang, 2014). Twitter: @jeanburgess

Andy Campbell is a critic-in-residence with the Core Program (Glassell School of Art / MFAH) and an instructor at Rice University. His work focuses on the interplay between contemporary art and historically marginalized sexual communities. Recent and forthcoming publications focus on womyn's lands and photography, as well as Afro-pessimism in the public artwork of Beverly Buchanan. His writings and reviews appear in *GLQ*, *Artforum*, *Aperture*, *Art Lies*, and other publications.

Nia I. Cantey is a former associate professor of social work. Currently, she is the independent living program manager with the Georgia Division of Families and Children Services. She has published peer-reviewed articles and book chapters on issues related to historically Black colleges and universities, gender and sexuality studies, and social injustices. E-mail: drcantey@gmail.com; Twitter: @Nowatj

Meredith Clark is an assistant professor in digital and print news at the Mayborn School of Journalism at the University of North Texas in Denton. Her work centers on representations of race in digital, social, and print media; her research addresses topics including online community formation, social identity maintenance via social media, and multimedia journalism pedagogy. Twitter: @meredithclark

Daniel Faltesek is an assistant professor of social media at Oregon State University. His research integrates logistical and financial factors in the production of media interfaces. Connecting these factors crosses media theory with computational image analysis, social networks, and critical/cultural studies. Web: www.danielfaltesek.com

Yang Feng is an assistant professor in the Department of Communication Studies for the University of Virginia's College at Wise. Her primary research area focuses on advertising effects and cross-cultural communication, which advances and applies prominent advertising and communication theories in the interrelated contexts of culture, politics, and communication technologies. E-mail: yf8f@uvawise.edu

Anne Galloway is senior lecturer in the School of Design, Victoria University of Wellington (NZ). As a design ethnographer, her research explores livestock farming practices and human-animal-machine ecologies. Anne lives in rural New Zealand with a human, a cat, four sheep, and many machines. Blog: http://morethanhumanlab.org; Twitter: @annegalloway

Jaigris Hodson is an assistant professor in professional communication and director of the Launch Zone at Ryerson University, in Toronto, Canada. Her interdisciplinary work examines social media discourses and the use of technology within organizations and between organizations and their publics. Web: www.jaigrishodson.com

Narayanan Iyer is a clinical associate professor in the Edward R. Murrow College of Communication at Washington State University Vancouver. His research work uses a social scientific perspective to study the processes and effects of mediated content. His research interests are in the areas of pharmaceutical advertising and digital communication.

Jenny Ungbha Korn is a scholar of race, gender, and online identity, with academic training in communication, sociology, theater, public policy, and gender studies from Princeton, Harvard, Northwestern, and UIC. Her work has been published in *Contexts*, *The Encyclopedia of Asian American Culture*, *Feminist*

Media Studies; The Intersectional Internet; The Journal of Economics and Statistics; Multicultural America; Our Voices; Television, Social Media, and Fan Culture; and Harvard's *Transition*. E-mail: KornPublic@comcast.net; Web: http://jennykorn.com; Twitter: @JennyKorn

Benjamin Lyons is a Ph.D. candidate and doctoral fellow in mass communication and media arts at Southern Illinois University Carbondale. His work in political communication focuses on social identity, discussion networks, contested beliefs, and fact-checking. His research also addresses the effects of new media forms in the broader deliberative system. E-mail: lyonsb@siu.edu

Mélanie Millette is professeure substitut at the Département de communication sociale et publique, UQÀM, Canada. Her work centers on social, political, and cultural aspects of social media use and how users mobilize online platforms to achieve political participation and get recognition. She won a SSHRC Armand-Bombardier grant and a Trudeau Foundation scholarship for her thesis research, which examines media visibility options offered by online participation channels such as Twitter for Francophone minorities in Canada. E-mail: millette.melanie@uqam.ca

Magdalena Olszanowski is an artist and Ph.D. candidate in communication studies at Concordia University in Montreal, Canada. She is a senior research assistant at her supervisor Kim Sawchuk's *Mobile Media Lab*. Her scholarly and artistic work on gender, technology, self-imaging, disability, and mobile image-based media has been published and presented internationally. Her dissertation is focused on the feminist online media histories of the 1990s. Web: http://raisecain.net; Twitter: @raisecain

Chang Sup Park is an assistant professor in mass communications at Bloomsburg University of Pennsylvania. His work centers on the study of social media, journalism, and politics, and his research addresses topics such as citizen journalism, digital journalism, political participation, civic engagement, information processing, online speech, and media policy. E-mail: cpark@bloomu.edu

Andrew Peck is a Ph.D. candidate in the Department of Communication Arts at the University of Wisconsin–Madison. His research focuses on vernacular discourse online and often engages with the rhetorical potential of digital humor and play, including Internet memes and the Slender Man. His work has appeared in the *International Journal of Communication* and the *Journal of American Folklore*. E-mail: ampeck12@gmail.com

Nathan Rambukkana is an assistant professor in communication studies at Wilfrid Laurier University, in Waterloo, Canada. His work centers on the study of discourse, politics, and identities, and his research addresses topics such as hashtag publics, digital intimacies, intimate privilege, and non/monogamy in the public sphere. He is the author of the book *Fraught Intimacies: Non/Monogamy in the Public Sphere*. Blog: www.complexsingularities.net; E-mail: nrambukkana@wlu.ca; Twitter: @n_rambukkana

Cara Robinson is an assistant professor in the urban studies program at Tennessee State University. Dr. Robinson currently teaches undergraduate courses in urban studies and nonprofit management within the College of Public Service and Urban Affairs. She is the coordinator of the TSU Community Democracy Initiative. Dr. Robinson has published research in a variety of public policy areas primarily focused on issues related to poverty, homelessness, and social equity. E-mail: crobin22@Tnstate.edu; Twitter: @cara_rob

Sylvain Rocheleau is a Ph.D. student in cognitive informatics at Téluq and UQÀM in Montreal, Canada. His work focuses on media and social media content analysis using digital methods such as web scraping and data mining on large-scale data sets. He is a research assistant at the Centre de recherche interuniversitaire sur la communication, l'information et la société (CRICIS) and co-founder of the Information Flow Observatory. E-mail : info@sylvain-rocheleau.com

Yoonmo Sang earned his Ph.D. from the University of Texas at Austin. In January 2016, he will start his new position as Assistant Professor at Howard University in Washington, DC. He studies the intersection of new communication technologies and the law. His scholarly writing has appeared in the *Journal of Media Law and Ethics*, the *Journal of Medical Systems*, *Computers in Human Behavior*, *American Behavioral Scientist*, and *Telematics and Informatics*, among others.

Anthony Santoro is an assistant professor of American history and American studies at Sogang University in Seoul, South Korea. His work focuses on American religious, sport, and legal history, particularly on professional football and on capital punishment. He is associate editor of *Religion and the Marketplace in the United States* (Oxford, 2015) and the author of *Exile and Embrace: Contemporary Religious Discourse on the Death Penalty* (Northeastern, 2013). E-mail: asantoro@sogang.ac.kr

Theresa Sauter is a research associate in the ARC Centre of Excellence for Creative Industries and Innovation (http://www.cci.edu.au/) at Queensland University of Technology in Brisbane, Australia. Her work considers technologies as techniques of self and explores the historical contingencies of processes of subjectivation. Her research interests include digital sociology, self-formation, socio-technical interactions, and practices of governing self and others.

Amanda Grace Sikarskie is Visiting Assistant Professor of Design History at Kendall College of Art & Design in Grand Rapids, Michigan. She is a historian of textiles, fashion, and needlework, with a particular interest in the intersection of social media and textiles. Her first book, *Textile Collections: Preservation, Access, Curation and Interpretation in a Digital Age*, is forthcoming from Rowman & Littlefield. E-mail: amandasikarskie@ferris.edu

sava saheli singh is a Ph.D. candidate, program assistant, and adjunct instructor in the Educational Communication and Technology Program at New York University. She is completing her dissertation on the role of Twitter in academic communities and scholarly communication, and is interested in interrogating techno-utopianism to highlight who gets left out. She is also a member of the editorial collective for the *Journal of Interactive Technology and Pedagogy*. E-mail: sava@nyu.edu; Twitter: @savasavasava

Aaron Veenstra is an associate professor in the School of Journalism at Southern Illinois University Carbondale. His research focuses on the intersection between social identity and communication, with a particular focus on online political and in-group communication, and social influences on cognition. He is the founder of SIUC's New Media Study Group. E-mail: asveenstra@siu.edu

Wenjing Xie is an assistant professor in the School of Journalism at Southern Illinois University Carbondale. Her research interests include communication technology, social media, mobile phones, media psychology, emotion, and privacy. Specifically, her research revolves around user experience with new technologies and the social and psychological effects of social media and mobile communication. Her work has appeared in *Communication Research*, *Computers in Human Behavior*, and *Journal of Health Communication*, among others.

Weiai Wayne Xu is a researcher specializing in social network and social media communication. He received his Ph.D. from the Department of Communication at the State University of New York at Buffalo. He is currently a postdoctoral research and teaching associate in the Department of Communication Studies at Northeastern University. He has published studies that examine online communities involving various communication scenes, such as knowledge sharing, political activism, cultural diffusion, and gatekeeping.

Joana Ziller is an adjunct professor in the Social Communication Department at the Universidade Federal de Minas Gerais, Brazil. Her undergraduate degree was in social communication and her PhD in information science. Her work centers on the relationship between common people and the media, especially ordinary people's audiovisual production. She is a member of the New Media Convergence Center. Her research related to the topic of this paper is supported by CNPq and Fapemig. E-mail: joana.ziller@gmail.com

Index

General Editor: *Steve Jones*

Digital Formations is the best source for critical, well-written books about digital technologies and modern life. Books in the series break new ground by emphasizing multiple methodological and theoretical approaches to deeply probe the formation and reformation of lived experience as it is refracted through digital interaction. Each volume in **Digital Formations** pushes forward our understanding of the intersections, and corresponding implications, between digital technologies and everyday life. The series examines broad issues in realms such as digital culture, electronic commerce, law, politics and governance, gender, the Internet, race, art, health and medicine, and education. The series emphasizes critical studies in the context of emergent and existing digital technologies.

Other recent titles include:

Felicia Wu Song
   Virtual Communities: Bowling Alone, Online Together

Edited by Sharon Kleinman
   The Culture of Efficiency: Technology in Everyday Life

Edward Lee Lamoureux, Steven L. Baron, & Claire Stewart
   Intellectual Property Law and Interactive Media: Free for a Fee

Edited by Adrienne Russell & Nabil Echchaibi
   International Blogging: Identity, Politics and Networked Publics

Edited by Don Heider
   Living Virtually: Researching New Worlds

Edited by Judith Burnett, Peter Senker & Kathy Walker
   The Myths of Technology: Innovation and Inequality

Edited by Knut Lundby
   Digital Storytelling, Mediatized Stories: Self-representations in New Media

Theresa M. Senft
   Camgirls: Celebrity and Community in the Age of Social Networks

Edited by Chris Paterson & David Domingo
   Making Online News: The Ethnography of New Media Production

To order other books in this series please contact our Customer Service Department:
   (800) 770-LANG (within the US)
   (212) 647-7706 (outside the US)
   (212) 647-7707 FAX

To find out more about the series or browse a full list of titles, please visit our website:
   WWW.PETERLANG.COM